The Frayed Atlantic Edge

The Frayed Atlantic Edge

A Historian's Journey
from Shetland to the Channel

DAVID GANGE

**WILLIAM
COLLINS**

William Collins
An imprint of HarperCollins*Publishers*
1 London Bridge Street
London SE1 9GF

WilliamCollinsBooks.com

First published in Great Britain in 2019 by William Collins

2

A catalogue record for this book is available from the British Library

ISBN 978-0-00-822511-7

Maps by Martin Brown

Typeset in Granjon by Palimpsest Book Production Ltd, Falkirk, Stirlingshire

Printed and bound in Great Britain by CPI Group (UK) Ltd, Croydon CR0 4YY

For Llinos, who taught me to love big seas
and small languages

CONTENTS

PREFACE

THIS JOURNEY INVOLVED arriving, dripping and bedraggled, in dozens of coastal communities. When I set out, I hadn't imagined just how generous the people whose homes and workplaces I dampened would be: without such openness, particularly evident on small islands, this project would never have got far. I learned as much through long evenings of discussion as through the other three resources on which the book is based: libraries, archives and the observation of land and sea from the kayak. It wasn't just the spectacles of sea cliffs, nor the dramas of ocean weather, but also those social occasions that meant I ended the journey with greatly intensified enthusiasm for scattered Atlantic islands like Foula, Barraigh and Thoraí.

Such conversations worked to strengthen the conviction I set out with: that British and Irish histories are usually written inside out, perpetuating the misconception that today's land-bound geographies have existed forever. Despite the efforts of authors such as Barry Cunliffe, whose *Facing the Ocean: The Atlantic and its Peoples, 8000 BC to AD 1500* (2001) inspired much debate among historians, the significance of coasts is consistently underestimated, and the potential of small boats as tools to make sense of their histories is rarely explored.

This book sets out to put some of that imbalance right, showing not only that Atlantic geographies have been crucial to British and Irish life but that they continue to be so. It is structured by region, because part of its purpose is to show how similar ingredients of wind, waves and rock have been transformed into entirely different island and coastal cultures by the divergent processes of history. The chapters were written in order, while I travelled, so my process of learning runs in parallel to the reader's experience of moving through the book: burrowing gradually deeper into the many ways in which the shorelines are significant. This allows the narrative to follow a trajectory in which the opening chapters evoke the act of kayaking, establishing sounds, smells, sights and stories of the venerable tradition of travelling at sea level. Only gradually does the balance shift towards historical research, literary criticism and argument, revealing the implications of new perspectives picked up through slow travel.

The final section, 'The View from the Sea', completes that transition. It switches to a different register as it unpicks historical significance from the chapters. It argues that the whole shape of British history is transformed by granting Atlantic coasts and islands a central rather than marginal role. The implications of key historical moments are problematised or reversed. The so-called Enlightenment, for instance, might best be interpreted as the triumph of a few cities – Dublin, Edinburgh, London, Birmingham – at the expense of other regions. For coastal communities it was the beginning, and the cause, of a lengthy dark age. In contrast, much of what were once referred to as Dark Ages had been eras of great coastal strength and enlightenment, when the intellectual traditions of the Irish Atlantic were the most advanced in Europe. Such reversals abound. The widely celebrated Education Acts of 1870 and 1872 were unmitigated disasters for many coastal zones, while the grim economic recession of the 1970s saw an island renaissance unprecedented for two centuries. All British history looks different when inland cities are made remote by seeing them from Atlantic shorelines, and the most powerful element of a year's journey by kayak was immersion in that changed perspective.

As this suggests, it's not just historical narratives but also familiar geographies that these waters erode. I began the journey believing I was travelling down a western edge of Britain and Ireland, and assuming I knew what that implied. But these Atlantic shores were long connected as closely to Reykjavik, Bilbao, or the Moroccan port of Safi as to London. There are echoes of Belize on Orkney shores and Nigerian history laps the coast of County Mayo. Communities at the edges were interlinked. Semi-detached from their land masses, they belonged to ocean. This is evident in the artefacts archaeologists unearth and the stories of shoreline encounters. Rare are the coastal regions that haven't woven the Armada into their folklore or looked to Scandinavia for ocean-going expertise. Scarcer still are the regions that didn't gain or lose from imperial encounters. These connections are just as clear in the foliage of British and Irish cliffs as in the records of trade or warfare: unfamiliar plants I sat or slept among often turned out to be a misplaced Spanish saxifrage or Norwegian liverwort.

Lerwick (Shetland) and Kinsale (County Cork) have been absent from London-centred histories of the British and Irish isles not because they lacked significance but because they operated in other geographic frames from Gravesend, Grimsby or Dublin: they saw different migrations of people, animals, goods and spores and seeds. At its most extreme, this phenomenon means that some sites on the Irish Atlantic reveal more evidence of historic sea links to China than to England. As I travelled, I found my preconceptions about Britain crumbling, destabilised from within by the diversity of coastal regions and from without by the stories shared by vast Atlantic littorals. Although the local specificities predominate in much of this book (being usually more obvious from a kayak) it was the moments when immense Atlantic geographies intruded that did most to challenge my mental landscape: one purpose of the final chapter is to bring these issues into focus, exploring what visions of the British Isles might emerge at their shorelines.

Just as the most significant histories often happen on the edges of the islands, the most interesting phenomena regularly occur in the margins between disciplines. Exploring past lives on coastlines meant reaching for ideas from geologists, ecologists, naturalists, geographers, anthropologists, artists, poets, novelists or musicians more often than historians. Seabirds, fish and species of seaweed play roles as significant in this book as politicians or their institutions: they had as great an effect on past shoreline lives, and the importance of island pasts today almost always relates both to ecology and community. Talking to naturalists, ecologists, archaeologists and artists was a highlight of the process of researching this book and I'd love to think that such lines of communication might one day be wedged more permanently open.

These ideas are the big themes reserved for the end of the book: conclusions drawn from the stories of exploring these phenomenal coastlines by kayak. That exploration – the biggest adventure I've ever undertaken – predominates for the next eleven chapters. Paddling beneath huge cliffs and across racing tides produced material that suits media other than prose. Though I didn't dare risk any expensive equipment, I carried a small camera to sea with me in order to take the photographs in this book. But, especially at the start of the journey, I took thousands of pictures. There is therefore a web resource to accompany the project at www.frayedatlanticedge. com. That site includes a photographic record to accompany each chapter, one or two short films, and further practical information for anyone wishing to paddle or research these coasts. It also contains links to the scholarly articles in which I explored the reasoning behind the project, and hosts an extensive bibliography. It is hoped that any reader who, after reading this book, seeks further immersion in the Atlantic waters of Britain and Ireland will find something of interest there.

INTRODUCTION

A Journey in the Making

I REMEMBER CLEARLY the moment I decided to embark on this journey. I'd shaken myself awake from a miserable night. The sun was yet to rise, but the view to the east was already full of promise. With overnight rain departed, a band of rich gold separated dark blue sky from the black silhouettes of mountains. The purring of curlews had begun to restore a sense of warm, active life to this cold, damp world and fulmars were wheeling over the water as the last of the rough night's swell died away.

The previous afternoon, I'd kayaked to one of my favourite places: Eilean a' Chlèirich. This was my last night outdoors for some time, and although a short squall was forecast, I felt the need to venture somewhere memorable. Eilean a' Chlèirich means 'Priest Island'. This single square mile of rock is uninhabited, and hemmed in by cliffs and boulders which prevent even small yachts from landing. Its upper slopes are home to storm petrels and other creatures that don't cohabit well with humans. The most remote of the Summer Isles off the north-west coast of Scotland, Chlèirich is a final, bleak, landfall before the Outer Hebrides.

Setting off from a small calm bay on the Coigach Peninsula, I'd made my way past the largest Summer Isles and along a chain of rocks that rise like wrecks from the sea. The wind had risen sharply as I battled waves on the final crossing and, with arms and thighs aching, it had been a great relief to reach water sheltered by the south-eastern cliffs of the isle. I clambered up the coast while pale November light gave way to storm clouds, and wandered above a

I

patchwork of tiny lochans to the island's northern point, where spray from a gathering swell soon rose higher than the cliffs. Because of the approaching wall of rain, I couldn't see the distant islands to the west, so I settled into my waterproof sleeping bag (figure 1.1), with pinkish sandstone boulders for shelter and my back to the weather. I was soon enmeshed in a drift net of wetness: salt-tasting rain seemed to enclose me from every compass point. My memory of those hours is defined by sweet smells of decaying island earth.

In the morning I blinked water from my eyelashes and stumbled to my feet. I was gazing downwards as I stood, carefully nudging the sleeping bag so the water on its shell didn't spill inside. Then I looked up, and the moment was heart-stopping. I must have turned around four or more times before I gained enough composure to choose a direction to look in. The storm had cleansed the skies so completely that every feature of the seascape was clear and perfect. A vast shattered coastline stretched on all sides: the tattered ocean-gouged fringe of northern Britain.

I was taken aback by the diversity of this view. To the west, the horizon was a long stuttering line of Outer Hebrides. The first rays of sun caught Harris' highest hill, An Clisham (An Cliseam); its silhouette, which should have been featureless at this distance, was bright with golden-brown glens and ridges. In the foreground, the Shiant Isles, puffin-covered in summer, rose like great bronze whale-backs from the sea. And above the northernmost point of the chain of islands was a stretch of blank horizon that marked open sea till Iceland. To the north-east, the coastline ran towards Cape Wrath, but as the mainland reached its terminus the land refused to give way: some of the weirdest peaks imaginable – Stac Pollaidh, Suilven (Sùilebheinn), Quinag (A' Chuinneag), Foinaven (Foinne Bheinn) – erupt like deformed molars on a vast fossil jawbone. These strange corroded towers were once sandbanks in a huge riverbed when this region was on the opposite side of the globe from the rest of Britain's land mass. There are many miles between each peak – long winding drives along narrow one-track roads – but the view from this spot concertinaed them together. The mountains to the south-east are less disorienting: where the northern peaks such as Suilven ('the Pillar') and Quinag ('the Milk Churn') challenge every preconception of what a mountain is, the hills to the south, such as An Teallach ('the Forge'), epitomise the pointed peaks and sweeping ridges a child might draw. These tips stand out from a skyline stretching via the magical Torridon range to the Isle of Skye in the south.

Although I've stood at 10,000 feet on peaks in the French, Swiss and Japanese Alps, the vistas from the rough knuckle at the centre of this tiny islet felt like the most expansive I've known. The British Isles are undoubtedly diminutive, yet this magical morning made me realise that how small they really are depends on how you measure them. The straight-line distance from Land's End in the south to John O'Groats in the north is just 603 miles (shorter than some roads in a state such as Texas or Ontario). Yet the first hundred miles of longitude on the mainland's north-west coast hold thousands of miles of coastline, with mountains, bays, estuaries, cliffs and islets that would repay a lifetime's exploration. Looking from Chlèirich at hills I'd climbed and stretches of coast I'd kayaked showed me that all I knew from two decades of wandering was mere fragments of something huge. I wondered what it would take to change that, and it was in that moment that the need to undertake this journey was born.

A few hours later I stopped in the port town of Ullapool. My mind had raced all morning as I tried to work out whether the plan I'd hatched could work. I headed for the town's two bookshops and filled three bags with reading that might help me think this through: tales of travel, natural histories, poetry, and accounts of Highland and Island life. Then I sat in a café, overlooking the pier from which ferries embark for the Western Isles, and began to consider the realities of what I was dreaming up. The trip couldn't be continuous: with a little planning, I could arrange my life to free up two weeks of each month, but the rest would have to be spent fulfilling responsibilities back in the English Midlands. This discontinuity would have two distinct advantages. It could spread the journey across the seasons, revealing every facet of the turning year on these weather-ravaged coastlines. It would also allow me to equip myself to tackle each stretch in the ways that suit it best: where one month I'd sit low in the water and power my kayak through the waves, the next I could don crampons to cross snow-clad peaks, or fix ropes to rock and descend into networks of mines and caves.

Over brunch in Ullapool I used my phone to search for things that could help me. The journey would require a large expedition kayak (five feet longer than the one I'd used that morning) to handle rough seas and hold gear for several days (figure 1.2). But the broken landscapes of the far north also made me look for a boat I could carry. I found a two-kilogram packraft: an inflatable vessel that could sit at the bottom of my rucksack until asked to carry me across a loch or along a stretch of river. Travelling like this I could spend my nights on islets and peaks with sight lines to the ocean and aim for 24/7 contact with the coastline.

Five hours later than intended I began the nine-hour drive south, but the sense of excitement was still building. Over the following months I renegotiated my life, striking deals and compromises to buy me time to travel. I rearranged my books so that the most accessible shelves in the house held only reading for this venture. I brushed up my learner's Welsh, and began to acquire a little Gaelic, so I'd have some access to more than just English writing on these coasts.

I mounted a two-metre-tall map on the wall of the room I work in and started to annotate its edge. I chose my starting place and date: Out Stack (a skerry north of Shetland) on 30 June. And I began to contact people who might help me on my way.

I'm a historian by profession: I teach courses and write books about nineteenth-century Britain. Like the work of many historians, my writing has focused, so far, on a few urban centres: it has done no justice to geographical diversity. I knew from past journeys that it would be hard to imagine places with histories, cultures and current conditions more different than, say, Shetland and the Isle of Barra, yet to many people these ocean-bound extremities might as well be interchangeable (and neither is likely even to be mentioned in a history book with 'Britain' in its title). This journey would be a quest to comprehend and articulate the intense particularity of the places on this coastline; in undertaking such a project I felt I could become a more rounded and responsible historian of the British Isles.

This is an especially significant task because the predominance of southern and central England enshrined in so much writing on Britain is a relatively recent development. It's not all that strange a fact, for instance, that in 1700 the island of St Kilda, now habitually presented as fiendishly remote, was among the most thoroughly documented rural communities in Europe. Metropolitan culture tends to take today's geography for granted, despite the fact that the British Isles were turned inside out by roads and rail. Mainland arteries – the Irish M8, the English M1 and even the West Coast Mainline – now run through the centres of their land mass rather than along the external sea roads that predominated till the railway boom of the 1830s. Since what would once have been miraculous – instantaneous communication across any earthly distance – has become ordinary, and what was once ordinary – travel by boat across a stretch of fierce sea – seems miraculous, attempts to empathise across centuries falter. Coasts and islands carry very different meanings than they once possessed: associations with remoteness and emptiness have replaced links with commerce and communication. This was part of the reason why travelling these coastlines felt like a way of thinking myself into the world of people I write and teach about.

But there were other reasons why this felt right. The belief that wandering the landscape is a productive technique for historical research is not unusual, or at least it didn't used to be. The links between historians and the outdoors were once strong. In the 1920s, for instance, G. M. Trevelyan wrote his classic histories of Britain while wandering Hadrian's Wall. Trevelyan soon became patron and champion of the many outdoors organisations that were all the rage after 1930. The links between tramping the countryside and doing history were still so clear in 1966 that when the Oxford historian Keith Thomas noted the rise of new kinds of scientific historian, he described 'the computer' replacing 'the stout boots' worn by 'advanced historians' of preceding decades.[1] Simon Schama wrote some of his best work in the 1990s, including a book called *Landscape and Memory*; at that time he frequently spoke of the 'archive of the feet'.

I discovered Trevelyan's writing in my teens, in a small Welsh bookshop on a family holiday, and learning about him was one of the things that set me on the trajectory towards my current life. At that time, part of me wished to work in the nearby national park, and part to write histories. Trevelyan made the two seem not just compatible but complementary. From that moment on, it was thinking of history as something that happened in negotiations between humans and hills, valleys, rain, wind and sea that drove me to be a historian. And I seem to have assumed from the beginning that reading and reflection are best done outdoors.

In those early years, while a pupil at the local comprehensive on the edge of the Peak District known as the Dark Peak, I'd wander past pubs and churches, new factories and old mills and onto the moors, where I'd try to memorise the physics formulae I needed for exams (only occasionally would short-eared owls or golden plovers distract me so much they'd write off a day's revision). My life over the two decades since then has been a quest for better ways to escape into the wild to think. From the modest moorland of the Peak District, to Scotland's least-peopled places and the hostile grandeur of Alpine ranges, my travels have extended and my attitudes to nature, work and literature become increasingly entwined. Now, whenever there's something I need to learn in detail I pack a bag with books and choose an atmospheric place to wander: I spend days over an unhurried journey and sit reading amid dramatic landscape. I've come to think that, with food and drink to spare, there are few luxuries more profound than getting well and truly lost for days among mountains. Staying still with a book for hours is also an excellent way to experience nature: a movement in the corner of the eye becomes a stoat between the boots; a sudden, startling noise is ptarmigan clattering onto nearby rocks; strange exhalations are a passing pod of porpoises. I have seen things, through this stillness, that I never would have otherwise: the most candid behaviour of otters and the preening habits of the little auk (figures 1.3 and 1.4). The associations this has created can be incongruous: Thomas Hardy and sea eagles, or Rebecca Solnit and long-tailed skuas. But it is this practice of reading, thinking and writing outdoors that has begun to hone the habits that make a year of journeying feel like the ultimate source of reflection and growth.

Many of the places this journey took me are now more free from human habitation than at any time since prehistory. The west has beautiful coastlines and wild ones, but even their remotest fragments are layered with diverse and difficult histories: they are sites of human default not design, shaped by past people but now reclaimed by nature. In the darkest spell of this story, the imperialism of nineteenth-century Lowlanders drove Highland and coastal communities inland, across the sea and to the grave. Part of the community of the island of St Kilda ended up in Melbourne, Australia; the people of Cork formed new Ontario communities; Welsh-speaking settlements were founded in the pampas and mountains of southern Argentina. The stories of these coastlines have stretched across the globe, revealing facets of Britain's imperial past and present very different from those seen from metropolitan London or Glasgow.

During my morning on Eilean a' Chlèirich I sought evidence of the people who once eked out livings in this most uncompromising spot. At first, wading through thick, ungrazed foliage, the island felt largely untouched. But I gradually began to see hints of human history shrouded by the plants: chunks of cut stone and roots of an old wall. The earliest human traces here are vestiges of stone circles from a time before written records: millennia over which imagination has freedom to roam. From a later age are scant remnants of Chlèirich's time as an early Christian retreat; this was the period that gave the island its name yet it is unrecorded in any document from the time. Then there are foundations from structures built by a nineteenth-century outlaw whose banishment from the mainland was recorded in just one short sentence of Gaelic prose. But the island's stones only really intersect with literary record with traces of the occupation by 1930s naturalists whose brief stay was immortalised in Frank Fraser Darling's *Island Years* (1940).

Barely anything of any of these people's endeavours stands above ankle height, yet Chlèirich is layered with past activity, where each successive wave of habitation has been so limited in scale that it hasn't erased previous histories. Wandering its hollows and hillocks is therefore a historian's or archaeologist's fantasy. Indeed, what made Chlèirich feel wild was not just wind, rain and the sounds of the sea, but the sense of being amid remnants of human action that had been conclusively defeated by weather. Humans toiled here centuries ago and my back when I slept had been laid against their labour: the rocks I nestled among had been worked by people, before wind, rain, ice and lichen reclaimed them for the wild. Although the British Isles have no untouched wilderness, their wildness is all the more remarkable for its entanglement with history: this journey would be an exercise in the art of interpreting the intertwining.

In that sense, my plan was an experiment. I hoped to see what could be learned by travelling slowly along these coastlines with an eye attuned to both the natural world and the remains of the past. The decades over which I've wandered here are long enough to begin to see changes and to ask what will become of these landscapes. The way in which some coastal regions were emptied of permanent populations now contrasts their growth as sites of leisure. Mountain paths grow wider and un-pathed regions fewer, coastal walking routes are extended and advertised in increasingly lavish brochures. I'd been spending nights on mountains for several years before I happened across someone doing the same, but now the experience isn't uncommon: in the winter before this journey I even slept on a Cairngorm summit from which the only visible artificial light was the pinprick of a head torch on a distant mountain. Thanks to social media and political devolution, communities from Applecross to

Anglesey pioneer new ways of living well while promoting and protecting the needs of nature. The languages of the small rural communities at the edges of the islands – particularly Welsh and Gaelic – grow in ways that once seemed impossible; lost languages like Norn have vocal advocates. 'Small language' networks of co-operation and exchange now link Cornwall and Wales with Breton and Galician cultures in ways that echo historic bonds along seaboards. Lynx might soon be restored to a few remote forests just as white-tailed eagles have been returned to seas and skies. Yet even the eagles are still a source of contention: beloved by tourists and naturalists they are resented, even sometimes poisoned and shot, by those who see them threaten livelihoods in farming, field sports or fishing. This book is therefore not just the story of a journey, or an exploration of past and present on the fringes of the British Isles, but a reflection on how far, and in what directions, our current interactions with the coast are reshaping this north-east Atlantic archipelago.

In attempting to tell this aspect of the story I wanted to rely on more than my own experience, so in the months leading up to my journey I made use of every professional and personal connection I had. I travelled to the University of the Highlands and Islands for events on coastal history, meeting, for the first time, the unofficial 'historian laureate' of Scottish coastal communities, Jim Hunter. I contacted artists and musicians, including the composer Sir Peter Maxwell Davies (an old friend of the family, who once taught me to play his Orkney-inspired music, but who passed away just weeks before my journey began). And I made use of my role as a teacher: I acquired dissertation students interested in the history and folklore of western Scotland, Wales and Ireland and wrote these places into my courses.

One class about these coasts was especially instructive. This was a seminar on 'Film and History' for the University of Birmingham's MA in Modern British Studies which I taught with a historian of the twentieth century, Matt Houlbrook. We chose early films of St Kilda and the North Sea as the case studies for our students. They began by watching the first moving picture of Britain's most famous small island: Oliver Pike's *St Kilda, Its People and Birds* (1908). Then they watched four films from the 1930s, including John Ritchie's footage of the evacuation, and Michael Powell's *The Edge of the World* (1937) which was set on Kilda but filmed in Shetland. We then chose three documentaries of the eastern coastline – John Grierson's *Drifters* (1929) and *Granton Trawler* (1934) as well as Henry Watt's *North Sea* (1939) – each of which places trawlers and fishing at its heart.

The effect of putting these films side by side is striking. They show the process of these coasts being mythologised. By the early twentieth century, the North Sea had come to stand for shipping, industry and progress: its early appearances on film were commissioned by the General Post Office to advertise the vibrancy of fishing fleets and the productive potential of the ocean. Trawlermen haul herring by the thousand from the waves: despite gales and storms, these icons of modern masculinity demonstrate human dominance over nature. Film-makers experiment with advanced techniques of sound and vision as they seek to portray the striving and struggling that make a modern factory of the sea. By contrast, the west in these films signals detachment and underdevelopment. Its communities hold out against terrible odds with only vestigial industries to aid them. A lone woman sits at a spinning wheel the same as the one her grandmother's grandmother used. A man is lowered from a cliff, draped in a sheet: he waits patiently, alone, to snag a guillemot which can then be salted for meagre winter sustenance. Children scatter, panicked by the strange sight of a camera and cameraman. Our students saw that when watching the east-coast trawlermen the viewer feels like the audience at a performance; when watching films of west-coast crofters and fisherfolk they were left with a feeling more like voyeurism.

The contrasts that appear in these films are fictions. They don't portray these places as they exist today nor as they were when the films were made; still less do they depict a world that could have been recognised in earlier ages. Yet stereotypes like these are repeated endlessly. Twenty-first-century poets are forced to work as hard as Norman MacCaig did in 1960 to remind readers that Gaelic verse is often small and formal: grandiose romanticism and the wild red-haired Gael live in lowland imaginations, not in west-coast glens and mountains. But I can't pretend that engrained romantic imagery doesn't still colour my own, lowlander's, obsession with these Atlantic fringes. Such notions are resilient to short spells on icy crags or a night in the ghostly remains of a cleared coastal township. But could they survive this journey's long immersion in these regions? I hoped to find my imagination changed by travel: the mists of Celtic twilight dispelled perhaps, with the delicate textures of mundane and everyday history appearing from the fog. This would not, I hoped, be a tale of disenchantment, but of changed enchantment, in which the rich worlds of real human beings exceeded (as any historian will say they always do) the hazy types of myth. So I knew, when I set out, what I wanted from this journey. But if journeys always turned out how we planned, and provided answers only to the questions we knew to ask, there'd be little point in taking them at all.

Shetland

— *Kayak route*

Orkney

Western
Isles

Far Northwest

Mountain Passage

Skye

ATLANTIC
OCEAN

Argyll &
Ulster

NORTH
SEA

Connacht

IRISH SEA

Munster

Wales

CELTIC SEA

Cornwall

ENGLISH CHANNEL

Let my fingers find
flaws and fissures in the face
of cliff and crag,
allowing feet to edge
along crack and ledge
storm and spume have scarred
for centuries
across the countenance of stacks.

Let me avoid
the gaze of guillemots,
the black-white judgements
of their wings;
foul mouths of fulmars;
cut and slash of razorbills;
gibes of gulls;
and let me keep my balance till
puffins pulse around me
and the glory of gannets
surrounding me like snow-clouds
ascendant in the air
gives me pause for wonder,
grants further cause for prayer.

Donald S. Murray, 'The Cragsman's Prayer' (2008)

SHETLAND
(July)

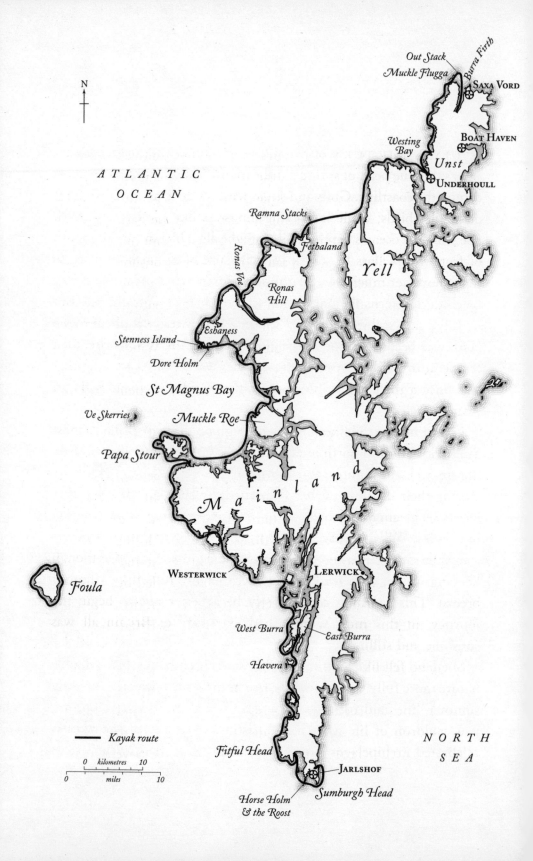

JULY IS THE TURN of Britain's year: counterintuitively, perhaps, it's the true peak of spring. At the month's onset, auks and waders throng the coastline. Gulls and skuas feast on the eggs and fledglings of smaller birds, while lumbering monsters like the basking shark rise from the ocean's depths to predate the algal bloom. In this month of frenzy, travellers by kayak can't be sure of an onshore place to sleep, however much they scrutinise the map: when a landing is met by chittering terns the only option is to slide back onto the sea. But by July's end, seabirds slip the leash that briefly tethered them to the land: wax becomes wane in the glut of coastal life. Winds rise, then temperatures fall, as species after species leaves, till every crag that was once a thick white fudge of feathers and excrement is flayed clean by gales.

I spent my first night on Shetland high on some of Britain's most dramatic cliffs and north of every road and home in the British Isles (figure 2.1). All night, seabirds returned to ledges below, gradually ceasing their daytime cackle; I watched the last light of a sun that barely set gleam on the backs of fulmars and puffins as they wheeled in to roost. When I woke (a mere three hours after closing my eyes) a fat skua sat feet away on the storm-stunted grass. It stared as though keeping watch, with feathers only occasionally ruffled by a hint of breeze. This morning could barely be a better one to begin my journey: in this most wind-lashed extremity of Britain all was sunshine and stillness.

Shetland felt like a fitting place to start. It embodies July's double nature more fully than anywhere else in the British Isles. In the early summer, 'the aald rock', as these islands are affectionately known, is a cauldron of life as rich and distinctive as any of the world's celebrated archipelagos, from the Galapagos to the Seychelles; its species – whether wrens, voles, moths or mosses – have evolved along unique trajectories. This month's journey will bid farewell to the fecundity of spring with a carnival of screeching, mewling life of which this morning's seabirds are just the start. The descent into winter in the Scottish mountains, when every plant or creature seems miraculous, will be dramatic.

Within an hour, early on the last day of June, I'll have paddled to Out Stack: a small rock that is the northernmost scrap of Britain. I'll turn. When I shift the sun from my right shoulder to my left, a journey that has filled my mind for months will begin. I wonder whether I should have some ritual ready: it'll feel odd for the act that begins this venture to be a paddle stroke like all the others. But I can't think of a ceremony that wouldn't seem ridiculous performed, alone, at sea. So I paddle north to my starting point, passing up a long, fjord-like voe called Burra Firth. This is lined to the east with Shetland's characteristic rich-red granite crags and stacks. To the west, a contorted, steely gneiss is shot through with quartz that, like the water, glints with silvery light. All the cliffs are swathed in a fleeting green: grass, moss and sea pinks cling to fissures in the rock through the short Shetland summer.

Reaching the mouth of Burra Firth was a decisive moment. If I turned right, around the red headland of Saxa Vord, I'd travel coasts sheltered from raw westerlies by the land mass of Britain. I'd write a book about the North Sea. But turning left is to choose the more austere Atlantic, its swell built through 2,000 miles of open ocean, and its coasts ravaged by some of the most powerful and unpredictable forces on the earth's surface. In her unparalleled trilogy of books on seashores, the Pennsylvanian Rachel Carson makes this coast a case study precisely because of the violence of waves which sometimes break, she says, with a force of two tons per square foot.[1] For now I was still shielded from swell by a long line of rocks, some with ominous names like 'Rumblings'. These outcrops are usually known simply by the name of the largest, Muckle Flugga, which is topped by a large, precarious Victorian lighthouse. Out Stack is the last and least imposing of the group.

Only later would I learn the need to ignore names like Rumblings and Out Stack, as late impositions on the landscape. It's a signal of Shetland's long separateness that the islands as their people know them are named differently from how they appear on maps: Out Stack, for instance, is merely a garbling of 'Otsta', a name still used by Shetland fishermen. These historic names of Shetland were collected and mapped for the first time in the 1970s, and those who undertook the task referred to the lived tradition they recorded as '100,000 echoes of our Viking past'. Muckle Flugga is among the names that reveal the resilience of local terms most clearly: for a century, officialdom imposed the bland 'North Unst' on this rock, but in 1964 gave in to the Shetlandic name which – derived from the Norse for large, steep island – speaks more eloquently of geography, history and Shetland's singularity.

Despite the shelter of the skerries, I proceeded south from Otsta with caution: as the sea spills round Britain's apex, strong tides can change a boat's course and sweep it into offshore waters. Just as the Atlantic breaks against these cliffs with unusual force, the tides round Shetland and Orkney are some of the most treacherous in the world. These forces, because they draw in floods of nutrients and prevent disturbance, are the skerries' greatest asset: they permit whales to feed and seabirds to breed.

On this still day, at the height of spring, this fecundity was spectacular. It felt like a stronghold: a vision, perhaps, of how all these shores might have been before human action ravaged them. By the time I left the firth, I was no longer alone but surrounded by life, and the new entourage that whirled around me provided the sense of occasion I'd thought impossible. A moment that could have been anticlimactic became entirely magical. A long string of gannets, slowly thickening, had begun to issue from the southern-most skerry of Muckle Flugga. Within minutes, hundreds of these huge birds – with wingspans of almost two metres – formed like a cyclone overhead. They circled clockwise, from ten to a hundred feet high, tracing a circuit perhaps a quarter-mile wide, each individual moving quickly from a speck in the distance to loom overhead (figure 2.2). Moments later, dozens of great skuas (known to Shetlanders as *bonxies*) joined the fray, pestering the gannets (*solans*) and drawing the only squawks from this otherwise voiceless flock. Black guillemots (*tysties*) and puffins (*nories*) flew by too, but took no part in the larger choreography, plotting small straight lines across the expanding circle.

More perhaps than any other bird, gannets evoke the bleak world of seaweed, guano, gales, crags and mackerel that sweeps north and west of the British Isles. Spending summer in dense communities, they colonise the steepest and most isolated elements of the Atlantic edge, building a world that looks like an oddly geometrical metropolis. Their chicks are known as squabs or guga, and dozens of these black-faced balls of silver fluff were visible on Muckle Flugga as I passed. During July the guga turn slowly black and leap from their ledges into a journey south that begins with a swim: they jump before they can fly. The young birds then make vast foraging flights, gradually securing a place on the edge of a colony that might be hundreds of miles from their birthplace. Then, they'll perch year after year in their tiny fiefdom, unmoved by everything the weather of Shetland, Faroe or Iceland can throw at them. I could feel no sense of identity with full-grown gannets, whose command of air and water transcends clumsy human seafaring; yet the guga's hare-brained, ill-prepared flop into the sea made me imagine it as an emblem of this journey's running jump into an alien ocean world. If I were ever to give my boat a name (and at least one Shetlander I met was taken aback, even offended, that I hadn't) I thought an excellent choice would be *Guga*.

Despite the infrequency of their squawks, the noise the gannets made as they swirled above was extraordinary. The sound of millions of feathers scything the air was enough to drown the ocean. This was the first time I'd considered the importance of hearing to the kayaker: unable to listen for dangers over the sound of the gannets, such as breakers over barriers in the sea, I felt shorn of a tool critical to navigation. And the thousand shadows of these powerful creatures created just the slightest sense of threat. Indeed, besides a few seventeenth- and eighteenth-century references to their sagacity and *storge* (the familial fondness they show towards their offspring), humans have rarely associated gannets with anything benign. Their appearances in art and literature are shaped by their most characteristic act: the fish-skewering dive from height into the depths. Wings folded back, the angelic, cruciform bird becomes a thrusting scalpel. This is, according to the leading naturalist's guide to the species, 'the heavyweight of the plunge-divers of the world' (and the gannet's evocative power is such that even this scientific monograph can't resist noting the bird's 'icy blue' stare).[2] In the 1930s, an island joke held that plans were afoot for the canning of 'fird' (gannets tasting like a cross of fish and bird) but that no tin could hold 'the internal violence from the northern isle': the gannet had come to stand for the storms of its northern outposts as well as its own oceanic stink and sudden plummet.[3] And the shift from soaring beauty to abrupt violence has long been a theme to build macabre visions on; as I moved beneath the avian storm cloud I couldn't keep the most sinister of gannet poems, Robin Robertson's 'The Law of the Island', from needling its way into my head. In this beautifully distilled poem, an island outlaw is lashed to a barely floating hunk of timber, with silver mackerel tied across his eyes and mouth. The islanders who have been his judge and jury push him into the tides:

> They stood then,
> smoking cigarettes
> and watching the sky,
> waiting for a gannet
> to read that flex of silver
> from a hundred feet up,
> close its wings
> and plummet-dive.[4]

This captures something of the force with which these bright birds, wreathed in shining bubbles, pierce the gloomy depths. Yet real gannets are ocean survivors, not kamikaze warriors, so there was no need for empathy with the island outlaw, and never a Hitchcockian threat in this great wheeling.

In fact, the leisurely hour I spent in the sun at Muckle Flugga would be the last moment of safety for some time. As I began the journey south down the island of Unst I hit a wall of breakers and swell that beat against the most preposterous cliffs I'd ever looked up at. With astonishing precision, fulmars traced the profiles of complex waves that seemed entirely unpredictable to me. Crests soon hit the boat from both sides, forcing its narrow bow beneath pirling water until its buoyancy saw it surge up through the foam. The bow would then smack down – diving through air where there had just been wave – into a sucking surface of receding sea. Twice in the first half-hour an unforeseen peak forced me sideways and into the ocean and I had to flick my hips to roll back upright, wrenching the paddle round to twist my body out from underwater (I was desperately glad of the previous week, spent practising short journeys in surf off North Uist with the most foolhardy kayaker I've ever met, my partner, Llinos – figure 2.3). As the last of my gannet escort returned to their pungent white promontories, I felt my sense of distance from everyone and everything keenly. I wouldn't see another human today, not even a silhouette on the cliffs that tower above. Even if someone was looking down, the roiling stretch of intervening ocean meant we might as well have been a world apart.

Passing down Unst was the hardest day's travel I'd ever done. In the evening I pulled into the shelter of a small cove, Westing Bay, with the sensation that I'd walked repeatedly through a brine car wash. I set out my sleeping bag on an islet called Brough Holm which, like so many tiny Shetland skerries, has a ruin attesting to productive purpose long ago. Covered in golden lichen, the remnants of this *böd* (fishing store) stand among deep-yellow bird's-foot trefoil which gives way suddenly to kelp and bladderwrack: a colourful world of greens, gold and brown that was made still richer by the evening light. The remnants of the Iron Age and Viking sites of Underhoull commanded the landward horizon, with a vantage along tomorrow's path, which would take me across a major sea road of the Norse world: the sound that separates the island of Unst from its southern neighbour, Yell, was once the easiest route between Norway and conquest.

Safe from the sea, I shuddered at the thought of what today's journey would have been like in less forgiving weather. I spent sunset drying out while reading about the small boats of Shetland, and thinking of centuries of families who'd rowed these coasts in all conditions.[5] Far from an anticlimax, this dramatic day felt like a grand fanfare to see me on my way. Although it would be a while before I learned to sleep well in July's perpetual light, I did doze for more than three hours that night, mostly unbothered by the outraged squeak of an oystercatcher each time a gull strayed close.

By some kind of miracle, the calm weather in which I set out held for days, with only brief early-morning interludes of cloud and

breeze. I was able to travel what should have been the most challenging stage of my journey with few hardships beyond some
sunburn round the ears. The two rolls in the maelstrom round north
Unst were my only submarine adventures. Covering an average of
thirty-two miles a day – not as the crow flies, but in and out of
gorgeous inlets with imposing headlands – I still had hours to read
or hang around at sea when gannets dived or porpoise fins rolled
above the waves. In the orange evenings and white mornings I
stretched my legs across the islands I'd chosen to sleep on and nosed
round their ruins (I've never been anywhere with so many abandoned buildings from so many centuries). I began to think up
questions for present-day islanders and for the past Shetlanders
whose lives persist in the archives. But this still idyll, I had to remind
myself, could not last.

The sensible way to undertake a journey along Britain's Atlantic
coast would have been from south to north. With prevailing sou'westerlies at my back I would have been working with, rather than
against, the weather. But I couldn't bring myself to do that. While
planning this trip in moments snatched from university teaching,
familiar English and Welsh coastlines felt like the wrong kind of
start. If I was to make sense of the Atlantic coastline, I had to begin
by disorienting myself with total immersion in the seascapes and
histories of a place I still knew mainly through clichés of longboats,
horned helmets, sea mist and gales. This place is the seam between
the Atlantic and North Sea, where waves rule Britannia and always
have. It is a coast of staggering diversity as well as a thriving cultural
hub: those coasts and that culture are thoroughly intermixed.

The bond between Shetlanders and their extraordinary tradition of small boats is rightly renowned. There are many Shetland dialect poems whose message boils down to the principle that a boat is more than a means of transport:

> Take time; name dy boat weel,
> fur du's
> naming a wye o life.
> Du's
> naming a attitude.[6]

Most families in nineteenth-century Shetland had a 'fourareen': a small, wooden vessel for inshore fishing and ferrying supplies (known as 'flitting'). Shetland's 'national poet', Thomas Alexander Robertson, who wrote under the name Vagaland, popularised an old Faroese proverb to sum up the ethos of this family vessel. This is now well known across Shetland:

> Fragments of battered timber:
> teak, larch, enduring oak,
> but from them may be fashioned
> keel, hassen, routh and stroke.

> A homely vessel maybe,
> we build as best we can,
> to take us out of bondage:
> bound is the boatless man.

Vagaland was born in 1909 in Westerwick, a tiny village on the Atlantic coast. Around this settlement are impressive *drongs* (the Shetland term for sea stacks). These are tall needles and prickly ridges, forming cauldrons in which the incoming Atlantic beats and swirls. Vagaland had reason to hate the sea: his father drowned here before young 'Tammy Alex' was a year old. But, like so many Shetlanders, he found poetry in boats, coasts, and rows or walks along the cliffs of the 'wast' side. Vagaland's verse is full of evocations of small boats in driving gales on 'da wastern waves', of constellations reflected in still seas, and of rhythmic songs of sailors and fishermen.

Boats were essential to a family like Vagaland's because Shetland life and laws necessitated coastal and inter-island links. An inhabitant of tiny Out Skerries, for instance, had rights to flay the peat from the more fertile island of Whalsay: like the people of most small islands, Skerries folk would regularly 'flit da paet'. This didn't just imply a single journey, but weeks of seasonal back-and-forth for cutting and turning to prepare the fuel for use. Provision boats, postal boats, fishing boats and social boats negotiated tidal channels in everything but the fiercest storms: many routes I took, between islets and along coasts, were once widely travelled in those ways.

Elegant Shetland-style boats now rest onshore in coves of the Atlantic coast, some in use and others in decay. But their distinctive form has a long and illustrious pedigree. The famous Gokstad ship, excavated in Norway and dated to AD 850, was accompanied by two small vessels that differ little from later Shetland examples. From the seventeenth to the nineteenth centuries, parts were bought from Norway to be pieced together on the islands. These Nordic kits made light, narrow and double-ended vessels. They 'pranced' on the water; their gunwales (the top edges of a boat's sides) tapered before the bow and stern so the ship would flex and twist, dancing with the waves in ways that few boats can.

Yet each part of Shetland had different ocean-going needs, so the Shetland style developed local variations. From the beginning of the nineteenth century, three things accelerated this divergent evolution. First, the supply of Norwegian kits was interrupted by the Napoleonic Wars, giving Shetlanders new impetus to build for themselves. A generation later, the advent of steamers to the Scottish mainland allowed access to Scotland's oak and larch, reducing the reliance of this treeless archipelago on Norwegian pine. Between these two changes, the need for vessels more suited to Atlantic conditions became horribly clear: in June 1832, thirty-one boats were wrecked in a sudden storm that lasted five days. A hundred and five fishermen drowned.

Soon there were boatbuilders all over Shetland, experimenting with styles suited to local waters. Of all the islands, Fair Isle – twenty-five miles south of Shetland mainland and twenty-four north of Orkney – maintained Norwegian design features such as the narrow beam and short gunwales most faithfully. Fair Isle craftsmen could rely on tides and isolation to bring enough drift- and wreck-wood to construct much of their yoals. The most specialised and distinctively shaped parts of the boat, however, had to be recycled from old boats into new ones. This explains why Fair Isle vessels were conservative in form.

In the rest of Atlantic Shetland, lightness was slowly sacrificed for ocean-going heft. Adaptation was modest in the south, and more dramatic in the north. At Sumburgh Head (the mainland's southernmost headland) an extraordinary tidal splurge known as *da roost* provided excellent fishing, particularly for pollock (called *piltocks* by Shetlanders). Saithe boats stayed close to shore, but needed to hang on the edge of the tide, controlled by two skilled rowers, while two others ran lines through the racing sea. These boats were shallow and manoeuvrable; each could run *da roost* more than once a day.

A little further up the Atlantic coast, around the islands of East and West Burra, fishing grounds were sheltered, so there was no need for long, deep or beamy (wide) boats to carry large cargoes on heavy seas. Jetties were rare so boats were dragged up beaches to be kept in *noosts* (hollows in the ground). Short overland carries could help avoid tidal streams around these small islands and peninsulas.

The result was that lightness remained a priority even as boats widened and lost their prancing flex.

The seafaring traditions further north were different. The demands placed on Unst boats grew rapidly after the Napoleonic Wars, partly because of the new confidence and abilities acquired by seamen returning from war, but also due to growing international demand for white fish which swam in grounds so far offshore that, in the words of an eighteenth-century commentator, distance 'sink[s] the land'. Boatmen began to take the extravagant risks associated with travelling to the edge of the continental shelf and spending nights on the wild fishing grounds known as *da haaf*. With lines up to three miles long, bristling with a thousand hooks or more, they fished after rowing or sailing thirty to forty miles from home.

The boats that answered these demands were known as sixareens. They were as large and muscular as little wooden rowing boats can get. The width of Norwegian precursors was expanded dramatically, because the new, stupendously long and heavy lines would have dragged a yoal over. Space onboard was such that once lines were laid, fishermen would make a fire in the middle of the boat: they'd light pipes, brew tea, perhaps cook some of the herring caught for ling bait, and pass the time until they began the four-hour task of 'hailing the lines'. Yet lightness had not been entirely sacrificed: even the most robust sixareen could be carried by its crew of six. Twenty-five to thirty feet from stern to bow, the biggest of these boats was less than double the length of my eighteen-foot kayak. While kayaks are designed to pop back up if submerged and overturned, sixareens were open-topped and undecked. If tipped or swamped they were lost: no one has ever righted a sixareen at sea.

To attempt to comprehend the Shetland experience of the Atlantic I returned to Unst on the first day of really rough weather. By chance, I arrived at the Unst Boat Haven on the same day in July that, 135 years earlier, a storm took fifty lives. Most accounts from survivors of the 1881 storm were chilling but similar word-pictures of still waters turning quickly violent, so that 'the sea commenced to rain over us'. But one document was different: it asked searching questions about how the characteristics of Shetland sixareens had shaped the tragedy. This text was notes taken in a 1979 interview with Andrew and Danny Anderson who had rowed sixareens in their youth: their father had survived the 1881 storm, but their uncle was killed by it.

I asked a custodian of the Boat Haven, Robert Hughson, whether he knew any more about the family. He recalled Andrew, in his nineties, telling a tale of being caught in sea fog (*haa*) with his father. As they fished, none of the six on board mentioned the *haa* or said anything about navigation. But when they completed their tasks, the older men just set to rowing and cruised straight into their *noost* on Yell. Andrew had a long career as master mariner and captain of one of the first supertankers, but insisted that he never discovered what skill allowed his father's generation to navigate fog without so much as a compass. Such are the stories the sixareens inspire.

But Andrew wasn't entirely dewy-eyed for the previous generation of boatmen and boats. In the 1979 interview, he explained why he thought late nineteenth-century changes in sixareen construction magnified the storm's impact. The heaviest and longest sixareens, he claimed, had lost their key advantage: the lithesomeness of the small Shetland boat. It was impossibly gruelling to prevent these boats being caught side-on to a rough sea. And when a boat plunged into troughs between waves (or 'seas' as the brothers always call them), even the best crews lacked time and strength to turn its helm upwards and make it what they called 'sea loose' for the next barrage. Andrew and Danny detailed the actions required of boatmen during the phases of a wave; they described main swells forty feet high and the complex action of the intervening lesser swells, as well as how to deal with each. Their imagery is rich: boats reaching messy peaks were 'running through a sea of milk'. And they detailed the nature of the dangers: a sixareen could take a breaking sea filling it to the gunwales, but failing to clear the boat of water by the next such crest spelled ruin.

Andrew described how in a gale, every captain had a choice. They could raise the sail and run, risking ruin on skerries and being pushed into unsafe landings. On the night of 1881 almost all sixareens took this option because skippers knew they were too heavy to manoeuvre under oar: ten were wrecked. The alternative was what the brothers called 'laying to'. The smallest sixareen at sea on the night of the storm was the *Water Witch*, an older boat exactly the same length at the keel as my kayak. This was the only boat whose six oarsmen dared confront the weather. The crew fought a war of attrition with the storm, rowing solidly into the oncoming sea all night to keep the boat from ever yawing side-on to breaking swell. They won a battle with the winds that no other crew could have taken on, and they rowed safely into harbour next morning.

Lying in the path of deep depressions that sweep the Atlantic, Shetland regularly sees beautiful weather for a few short hours, sandwiched between storms and blanket fog: the crew of a sixareen rarely had the luxury of knowing what seas they'd confront. That they risked everything in small wooden craft for modest hauls of fish demonstrates their intrepidness, but also indicates the harsh conditions – both climatic and political – in which islanders often found themselves during times of oppressive governance before the present era of oil-driven affluence. As the Shetlander John Cumming put it,

The boat as transport and fishing tool has shaped so much of Shetland's history and its culture. We worked the land, true, but all too often a poor, thin land, and in desperate times fishing kept us alive. It made us who we are, a virile, confident, skilled and highly adaptable people with many stories to tell and a unique tongue in which to tell them.[7]

I had assumed, when reading all the admiring writing on the age of the sixareen, that the narrative of Shetland fishing must be one of decline from a golden era when saithe or ling could be plucked from the sea at will, so was surprised to find that the catches of Shetland ships have never been greater than they are today, the technological skill of earlier boatbuilders growing through the age of nets and herring (rather than lines and ling), then steam, motors and engine grease. The question of whether such innovation has been wholly positive – saving lives, but devastating sea life while reducing the number of livings to be made from the sea – is a different matter altogether.

In the serene conditions of the first few days, I felt distinctly un-intrepid, with Shetland's boating tradition a constant reminder of what humans are capable of when confronting the profound forces of the ocean. I moved slowly on, finding kayaking most pleasurable at night, when winds were lowest and the light dramatic. Between spells of paddling I worked through books I'd picked up in Lerwick, discovering more about the traditions and stories of these islands. Much of this reading was done in the boat, rocking gently on the swell as I rested. Without the splash of constantly rotating paddles, birds and animals often popped up close by, some reacting with surprise but others with curiosity at this bright intrusion in their midst. One gannet appeared within touching distance, allowing me immersion in the infamous ice-blue eye. Like a salt-rimed sea swan, it arched its wings defiantly, but made no sign of moving off. Seated on the waves, the bird's white tail feathers and black wing tips stretched a surprising distance from its bill (which, slightly hooked, resembled interlocking plates of some long-tarnished metal). Minutes later it launched itself past my bow, bouncing repeatedly on the water and coating my camera lens in sea spray: as elegant as a camel on ice. When I landed, I couldn't resist a look at the Shetland dialect 'wird book' I'd brought along, in case this maritime language had words to evoke things I'd been seeing. To my pleasure, sea spray was *brimmastyooch*.

It took three leisurely days to pass down Unst then Yell and reach the Shetland mainland. The crossing from Yell to the mainland was the most challenging hour since Unst. With tidal streams running along the sound, around the imposing Ramna Stacks to its south, and then across Fethaland (the finger thrusting out from north mainland), there was no possibility of tackling the whole crossing during the brief slack water between the incoming and outgoing tides. Today, the overnight cloud refused to lift quickly, and a few gusts from the *haaf* helped amplify my trepidation. But if I had one regret about my paddle down the difficult stretches of Unst and north Yell it was that I'd been too cautious in taking photos: however unsettling it proved to be, I resolved to use my camera even when among the contorted waters at the Stacks (figure 2.4).

The coast of north mainland – the region called Northmavine – is perhaps the most outlandish landscape in Britain: I'd entered a science-fiction vision of an ocean planet. The first headlands, gnarled, grey and viciously gouged by sea, contain some of the oldest rocks in Britain. These soon give way to young red granite pillars and pinnacles, topped with puffins or Arctic terns, which rise directly from the ocean (figure 2.5). Some are smooth and torpedo-like, others prodigiously spiked, and still others have broad bases cut through by arches resembling the galleries of a flooded cathedral. I thought of 'the living floriations and the leaping arches' of David Jones's long poem *The Anathemata*; in presenting cathedral architecture as an extension of the natural world, Jones insists that nature and culture shouldn't be seen as separate. Passing through in windless conditions gave me rare access to each dark transept in these enclosed, steep-sided spaces.

These islands are drowned mountains. Six hundred million years ago a vast 'Caledonian' range stretched from what is now Norway to the present-day United States. The islands of Shetland were then peaks of Himalayan majesty, before aeons of erosion ground them to their cores. This mountain heritage shapes Shetland's modern character: the ocean floor falls away from these 'erosional remnants' faster than from most of Britain, so that a depth is reached in half a mile that takes a hundred miles to reach from many English shores.[8] The behaviour of the ocean and the distributions of fish or oil are all defined by those underwater inclines. The first, grey headlands in Northmavine are a rare point at which no remnant of the Caledonian mountains survives, worn down to bedrock laid 3 billion years ago from quartz and feldspar, before laval heat deformed it into the coarse gneiss basement of today. Later, thick sediments settled over this foundation before the clash of continental plates which, through buckling, thrusting and folding, made the Caledonian mountains. When I gazed up at many mainland cliffs, I was staring through cross sections of those ancient hills, with an access to the distant past that is rarely possible from land. The vast variety of rocks – including granite, marble, limestone, gabbro and sandstone – generates the diversity of foliage above. Limestone feeds patches of green munificence, while gneiss and granite starve the ground into blanket bog.

Geological distinctiveness has drawn scientists and artists to Shetland for generations. The driving force behind the great twentieth-century renaissance of Scottish literature, Hugh MacDiarmid, moved here in search of 'elemental things', by which he meant old language as well

33

as rocks and the forces that moulded them. In ways that are often neglected, his career was defined by Shetland: it is indicative of the scale of Shetland's impact on his work that most of pages 385–1,035 in his collected poems were written here. MacDiarmid's son described the strange scene in their Shetland fisherman's cottage:

> The blazing peat fire, surviving in its grey ashes through the hollow of the night to be fanned fresh with the rising sun, patterned his legs to a tartan-red, and great blisters swelled. But nothing matched the white heat of passionate concentration, the marathon of sleepless nights and days that suddenly ended the sitting around for months indulging in that most deceptive of exercises – thinking.[9]

When MacDiarmid explored these islands – in his own words, 'rowing about on lonely waters; lying brooding on uninhabited islands' – he was actually in the company of a geologist (Thomas Robertson) and a fisherman (John Irvine of Saltness). It seems hard to imagine a group better suited to exploring Shetland than these experts in boats, rocks and words. One result of this alliance, MacDiarmid's long poem 'On a Raised Beach', is perhaps the finest evocation of the entanglement of Shetland's geology, sea and culture ever written. It becomes a kind of metamorphic metaphysical, with a famous opening – 'All is lithogenesis' – that rang true as I weaved my way through dramatic features formed from vertical layers of differentially eroded stone. This is a poem that demands to be read aloud, and deliberately snares the reader in thickets of dialect and science: words piled together like stones on the beach.

I dug my MacDiarmid from my dry bag and savoured his insistence that seeing is not enough when we confront Shetland rock:

> from optic to haptic and like a blind man run
> My fingers over you, arris by arris, burr by burr,

Slickensides, truité, rugas, foveloes
Bringing my aesthesis in vain to bear
An angle-titch to all you corrugations and coigns.[10]

When I reached Ronas Voe, one of the most dramatic sections of this coast, I took a kind of risk I never had before, sleeping at the bottom of tall red cliffs, where a tiny beach of silver-white stones seemed to stretch two yards behind the tide line: just enough space to keep me and my kayak out of ocean. From here, I looked up to Ronas Hill, the highest point in Shetland: a frozen former magma chamber from a huge Caledonian volcano.

Although the days were still fine, this was my first really wet night. Each day so far, I'd stopped kayaking a couple of hours before sleep in the hope of drying out. While at sea, I wear wetsuit boots, neoprene trousers and a thin rash vest that starts out black but after two or three days is mottled silver with salt. Each evening I'd change into warm and comfy land-wear. Sitting back in the sun, with a book by a Shetland author, I'd eat bread and cheese while watching the tides and seabirds pass. But tonight a thin drizzle set in, making the stones of my little beach shiny and slippery.

Rain makes decisions that would otherwise be of little consequence, such as when to change clothes or how to pack the sleeping bag, into significant moments when mistakes can lead to days of discomfort. I had just one set of land clothes (and nothing has ever dried in the hold of a kayak). So tonight, I rushed straight from my kayak gear into the sleeping bag and went to sleep early. Dampness increased as the rain thickened. True to form, my waterproof sleeping bag let nothing in for hours until I was woken by a thin trickle of water rolling down my neck and pooling above my collarbone. I was glad to be sheltered from the wind that had picked up, but decided that the best chance of comfort was to make a very early start.

This was the day I'd reach St Magnus Bay: a bowl in the side of Shetland, fifteen miles across, 140 metres deep and forty miles along the involuted edge I'd paddle. Many of the most dramatic remnants of Shetland's tumultuous geological past line its circumference. I decided to take my time over the first stage of this journey, passing down the stupendous Eshaness cliffs before landing in a cove at a tiny settlement called Stenness and walking to spend the night on the precipice I'd kayaked beneath. But conditions were slowly changing. Where the sea until today had been a blue-vaulted expanse with perpetual views, the swell had risen overnight to become a series of narrow corridors whose silver walls could obscure even the tallest cliffs. I'd wondered whether I might see basking sharks round Eshaness, but today they could have passed within feet without me knowing. For most of the day, this swell was immensely peaceful, its phases gentle and unthreatening as I moved through four dimensions with every stroke. But at the bottom of the Eshaness cliffs, the swell seemed to come from all directions at once: unpredictable and disorienting. The way to deal with this in a kayak is less about the arms than the hips and thighs. At leisure, I sit straight-ish, but the means of responding to complex seas is to lean forward, lower the centre of gravity and grip the kayak tightly with the knees, creating a sense of connection with the boat. After a damp night and in cold wind, I was deeply reluctant to roll this morning and relieved once I reached the simpler water near Stenness Island. Here I stopped on the swell to take pictures back along the cliffs, as best I could, and to observe the place I found myself: the bowl of another volcano. It felt strangely appropriate that in a spot where I was separated from writhing magma only by time, I should be plunging and leaping with violent swell, head rhythmically plunging beneath where my feet had been.

I landed at Stenness in time for lunch, left my kayak beside an old *haaf*-fishing station, and began the walk towards the lighthouse to learn these cliffs by watching the setting sunlight shift across them. Before I'd got beyond the bay, the local crofter approached and we discussed the caves along this coast. He insisted I explore the hollow interior of the island known as Dore Holm, rather than just looking up at its huge arch, and told me about a small entrance just north of the Eshaness lighthouse that had been found only a few years ago. It leads, he said, to a huge chamber inside the cliffs. This was my first reminder that apparently timeless and knowable cliffs are mysterious and shifting. There'd soon be many more.

Next day, I continued along the sweep of St Magnus Bay, spending a night that was close to perfection on the cliff-bound island of Muckle Roe. The cove I pulled into through a small gap in the island's towering crags was as rich in grasses, bog cotton and small flowering plants as anywhere I'd seen. Unsurprisingly, the products of the slow dissolution of contrasting rocks – a wildly varied sward of grasses and flowering plants – appear in Shetland culture almost as much as fish and boats. The star among young poets writing from Shetland today, Jen Hadfield, is particularly attentive to plants of bog and cliff such as the butterwort:

> I've fallen
> to my knees again not five
> minutes from home: first,
> the boss of Venusian leaves
> that look more like they docked
> than grew; a sappy nub;
> violet bell; the minaret
> of purpled bronze.[11]

In 'The Ambition' she even dreams of becoming butterwort, lugworm and trilobite, though her ultimate ambition is to be ocean, 'trussed on the rack of the swell':

The tide being out, I traipsed through dehydrated eelgrass
and the chopped warm salad of the shallows, and then
the Atlantic breached me part by part.[12]

I sat in this immersive scene and watched Arctic skuas (*skootie alan*) chase Arctic terns (*tirricks*) as, in displays of balletic brutality, they forced them to drop their catch or vomit recent meals. And as the air cooled, moths began to clamber up the grasses: after sleepily fumbling upwards they'd shift abruptly, as though at the flick of a switch, into a manic spiral through the evening air.

After Muckle Roe, a long voe leads far inland, ending in the town of Aith. Following the coast now meant plunging into the heart of mainland. Here, I visited Sally Huband – ecologist, nature writer and Shetland-bird surveyor – who made me soup and pizza, as well as providing valuable local knowledge for the next stages of my journey. Sally explained several of the characteristics of Shetland's wildlife that had struck me as I travelled. She told me, for instance, that the absence of peregrine falcons is partially explained by the dominance of fulmars in their favoured nest sites: when threatened, fulmars spit a thick oil that's enough to debilitate a peregrine chick or compromise an adult's flight. Sally had just flown back from the outlying island of Foula where she'd been collecting great-skua pellets as a favour for a friend who needed them before going to Greenland. Her descriptions of Foula's geological and biological distinctiveness convinced me that once I'd finished my month's journey south I would have to make my way there: without reaching Foula I couldn't claim to have travelled Atlantic Shetland.

Back on the water I headed for the mouth of St Magnus Bay, but before I could round the lower lip a large island blocked my route. For any sea kayaker, Papa Stour is the 'jewel in the crown' of Shetland: a mile-wide rock with a twenty-two-mile coastline, pocked with some of the deepest and most complex caves and arches in

Europe. It is stranded in the ocean amid speeding tides, and I decided to break the journey through them with a night high on the island's cliffs. I watched a trawler pass the remote Ve Skerries (another ancient knuckle of bedrock) as a group of Arctic skuas clucked and quarrelled like the drunk family at a seaside caravan park.

Papa Stour also appealed to Shetland's first geologists. In 1819, Samuel Hibbert described his arrival across transparent water, which made the boat appear 'suspended in mid-air over meadows of yellow, green, or red tangle, glistening with the white shells that clung to their fibres'. He observed 'red barren stacks of porphyry' that shot up from the water, 'scooped by the attrition of the sea into a hundred shapes'. Hibbert described many customs, including the Papa residents' tradition of trapping seals in the famous caves to club them until 'the walls of these gloomy recesses are stained with their blood'; but he gives a more picturesque vision of his own journey underground:

> The boat . . . entered a vault involved in gloom, when, turning an angle, the water began to glitter as if it contained in it different gems, and suddenly a burst of day-light broke in upon us, through an irregular opening at the top of the cave. This perforation, not more than twenty yards in its greatest dimensions, served to light up the entrance to a dark and vaulted den, through which the ripples of the swelling tide were, in their passage, converted by Echo, into low and distant murmurs.[13]

Hibbert was a polymath prone to taking long excursions in corduroy breeches and leather gaiters, accompanied by his dog (delightfully named Silly). It was one of these excursions that took him to Shetland in 1817, but his relationship with the islands was transformed when he happened across commercial quantities of chromite on Unst. In 1818 he began a geological survey covering all the archipelago. In the evening or during storms, he would appear at the doors of crofters, seeking bed or food. Then, according to his daughter-in-law, he would 'retire to rest lying down in his clothes, dry or wet, on a bed of heather or straw, but not always sleeping, for swarms of fleas might lay an interdict on sleep'. On Papa Stour his hosts treated him to tusk fish and 'cropping moggies' (spiced cod liver mixed with flour and boiled in the fish's stomach). Such dainties, he writes, should make Shetland a place of pilgrimages for discerning gourmands. He adds one caveat: for variety the poor islanders sometimes resort to coarse foodstuffs like lobster.

In the morning, I explored the caves, though I couldn't pass far into their depths, lacking the conditions that Hibbert recommends 'when the ocean shows no sterner wrinkles than are to be found on the surface of some sheltered lake'. I then swung round the headland beneath St Magnus Bay. Passing under yet more rugged cliffs, I called in on the memory of Vagaland at Westerwick, before embarking on the final stages of my Shetland voyage.

Once I was beyond the geological spectacles of the north, I made my way towards a world of small, fertile islands that were long smattered with settlements but are now home only to sheep. The day I set out through these islands was my first experience of the infamous Shetland *haa* and so the first time I really had to navigate. After ten minutes on the water, I tore my new compass from its plastic packaging and checked I could read it as I rocked. The conditions were haunting. Sometimes the *haa* sat flat against the gentle, three-feet swell. At others, it hung just inches from the water and tendrils of grey-white cloud seemed to stroke the surface of the waves. I had intended to spend the night on the island of Havera. But this, I'd since heard, was a place without streams: its community had been sustained by two wells that, a century later, I couldn't afford to rely on. So I aimed first to find my way between the Peerie Isles (*peerie* being Shetland dialect for small) to the outside tap at the local Outdoor Centre.

Mist is an excellent ally in wildlife watching. Today, not much after 4 a.m., an otter stood and watched as I drifted quietly by, while several red-throated divers, known here as rain geese, left it late to sidle off. Conditions, landscape, wildlife and atmosphere had all changed dramatically since St Magnus Bay. After landing and stocking up on water I set off for Havera, a place I'd long been intrigued to see. Slowly, the mist rose, wisps clinging to the moors east of the island, so that Havera gradually brightened from the west. Soon, it was stranded in a wedge of weak light beneath dark and silent skies. Clouds still licked the feet of the rough pewter cliffs long after their brows were clear. As I entered the mile of open water between mainland and the island, the swell was slow and gentle. On this day more than any other, the strange sensation of movement in multiple dimensions was something my body would retain: when I slept, many hours later, the cliffs seemed to undulate beneath me.

Havera is surrounded by richness. *Nories* and *tysties* surfaced laden with sand eels; forests of green-brown kelp seemed to grasp towards the surface. I'd planned this day and the next to give me time on the island. Here, if anywhere, I could begin to comprehend the lost communities of Shetland and the ruins that line the shores. The Havera folk left behind an archive, including recordings describing life on the island. A collaborative project of photography, research and writing used these to build a beautiful book, *Havera: The Story of an Island* (2013). I'd saved this to read *in situ*, where I could follow up each reference to a hill, promontory (*taing*) or rocky inlet (*geo*) by exploring it myself. Later, I'd spend a day on the Shetland mainland listening to recordings of the people of Havera and contextualising the extracts in the book. This was my best chance so far to explore the 'archive of the feet'.

Travelling south, I reached the island at the deep clefts of Stourli Geo and Brei Geo on its north side, and used a dialect poem, Christina de Luca's 'Mappin Havera', to guide me towards a landing. This poem begins with a warning:

> Havera's aa namit fae da sea.
> We could box da compass
> o wir isle; hits names markin
> ivery sklent da sea is med,
> da taings an stacks an gyos.

> If on your wye ta Havera
> an mist rowled in,
> ivery steekit bicht spelt danger;
> you had ta ken dem, ivery een.[14]

The poem then provides the necessary 'kenning', tracing the aids and obstacles encountered by fishermen at the island's edge, until spying safety at 'Nort Ham,/wir peerie *haven*/Mak for dere if you can'.

Having followed these directions, I pulled into Nort Ham at the island's south-east corner (figure 2.6) and was met with an onshore sea of wildflowers and grasses. I pulled my boat up among the buttercups, where I found the egg of a wheatear (*sten-shakker*), plundered by a neater and more precise predator than the *bonxies* who might ordinarily be culprits. But there were also *sten-shakker* fledglings, less cautious and more curious than their chattering parents. Without this pretty sheltered inlet at Nort Ham, Havera might never have been inhabited. Over centuries, the people and goods that entered or left the island came through this tiny gap: the community's single link to the world beyond.

I packed myself a bag of food and warm clothes, since new weather seemed to blow in by the hour. Havera has several landmarks I wanted to investigate. The most substantial is the abandoned village packed tightly into the corner known as Da Yard, which was the last place in Shetland's small isles to sustain a population. In 1911 it boasted twenty-nine occupants from five families. Inland, there are two outlying buildings. One is an old schoolhouse, the other the imposing trunk of a huge ruined windmill: a *meed* (sea mark) visible from many miles away. I decided to take my bearings according to the landscape before investigating the ruins. This proved trickier than expected since most headlands were colonies of terns: to disturb them is not just harmful to a threatened species, but also draws a volley of intense and committed dive-bombing far more likely to cause injury than the infamous attacks of great skuas (a *bonxie* attack is never conducted in such numbers or with such frenzied persistence).

When I finally found a spot to sit and read, I was gazing out onto Havera's satellite, West Skerry. Each spring, island sheep and cattle were swum here, across the hundred-yard sound, to protect the Havera crops. This ridge of rough pasture, completely separate from Havera's arable land, was key to the success of the island community. But it is arable richness that was Havera's greatest asset. The island is less than a square mile in size, yet its interior is not the rock- and wrack-strewn waste found on many islets of similar area. It's an unlikely idyll of well-drained, fertile earth, perpetually replenished by soft limestone that intrudes in veins through the region's granite bones. From the era of Neolithic field boundaries (still sometimes traceable), to the moment when the crofters left, this limestone made Havera a fine place to grow grain. Indeed, its name is probably derived from *hafr*, the Old Norse for 'oats'.

The island was generally presented by its last inhabitants as a plentiful and perfect home. Gideon Williamson died in 1999, seventy-six years after he left Havera; he remembered his birthplace as unique in Shetland because its fertile land was not 'just bits a patches here an dere' but one great expanse of rock-less, weed-less loam: 'you could tak a ploo an ploo da whole lot up . . . Hit wis entirely clean.' Shetland tradition accords to tilled Havera earth pest-repelling tendencies that verge on the magical.

Yet the topography did have drawbacks. Wells were no substitute for streams because running water had uses beyond cleaning and quenching thirst. The most common ruins I'd passed along other Shetland coasts were small, simple watermills built where rivers met the sea. But the people of Havera had to row their grain to Scalloway (a five-mile crossing) or Weisdale (eight miles) and pay for its grinding. This added labour, cost and risk to the challenges of island life. In the 1860s, a solution was dreamed up: the only windmill ever built in this storm-ravaged archipelago. The innovator might have been Gifford Laurenson, a skilled mason who was entrusted by the Society of Antiquaries with repairing the Iron Age Broch of Mousa (then 'mouldering into dust'). Between 1848–52, the Laurensons married into Havera families twice, and Gifford's sister and father (also a mason) moved to the island.

The significance of Gifford Laurenson's link to the Broch of Mousa is that the Havera mill evokes ancient Shetland more than modern. It stands like a round Iron Age edifice in a region where circular buildings are rare. It is a landmark that, like an ancient fort, puts Havera on the map: the most instantly recognisable of the small islands and, according to one local seafarer, 'a kinda lodestar for whaar you wir'. That incidental function is all well and good. But so many compromises were made with the mill's design, in order that it might withstand the Shetland weather, that it was useless for grinding grain: Havera folk quickly quit and resumed their mainland journeys.

The large, ill-fitting stones make the mill easy to climb, so I edged my way carefully up the green and golden lichen to an exceptional vantage on its walls. From here the world of which Havera was the centre could be surveyed. The rough low hills of the southern mainland, with their scattering of small white houses, occupied the eastern horizon, the shores becoming ever sandier as the hills sank and stretched south. These were the coastlines that Havera folk traversed on calm summer days. But beyond the southern extremities of the mainland, thirty miles distant, Fair Isle stood out on the horizon. From there, sweeping west, a long stretch of ocean was punctured only by the mad cliffs of Foula, frequently referred to as the wildest inhabited spot in the British Isles. These two islands signalled a remoter world to which, in certain weathers, Havera most certainly belonged.

I wandered downhill to the village, where small buildings are packed into a narrow isthmus with surprising neatness. This order is a product of the history of habitation. Although Havera had long been lived on (this is probably its first period of abandonment in millennia), the population was increased, and several new crofts built, during a sudden enforced settlement in the eighteenth century when landlords aimed to increase rents from every part of their domain. When the island was abandoned, it wasn't because of clearance, but was an after-effect of the overpopulation engineered by the lairds. A few Gaelic place names round the coast suggest that some incomers might have been Scots, not Shetlanders; two very different cultures forced together.

The overwhelming impression amid the ruins was of how communal life must have been. Doors and windows of the small neat houses look onto one another, while amenities such as the single village kiln suggest prominent shared spaces (in most of Scotland at this time each croft had a kiln of its own). Inside houses, outlines of two rooms, the *but* and *ben* (living room and bedroom), are often clear, although the interiors of some have been converted into well-crafted winter sheep pens. A single habitable house stands on the inland edge of the village. It is used by those who tend the island's sheep but its occupation seems irregular: a faded chess set and 1990s magazines accompany a toy animal that stares incongruously from a window.

This building shows that the village is not entirely abandoned. But nor is it entirely uninhabited. As I walked around its eastern edge I realised that the honking of fulmars was not just coming from the precipice below, but also from the ruins of crofts: the birds use the village walls as crags. Despite the healthy human population shown in the 1911 census, the last residents left the island in 1923. Jessie Goodlad, born on Havera in 1903, explained why:

> Dey left becaas dey wir naeboady left ta geeng back an fore wi da boat . . . becaas da young menwis aa laevin . . . weel, dey wirna laevin exactly, dey wir aa gyaain tae da fishin . . . dey wir naeboady to steer dis boat, dis saily-boat, back an fore.[15]

Fulmars spread across Britain's Atlantic coasts in the early decades of the twentieth century. The first of these birds may well have begun to nest here after the Havera folk had left (there are still retired Shetland fishermen who recall the time they were told, as children, to come and see this strange bird, the *maalie*). There is as much social change in nature, and as little permanence, as there is among people. Few events demonstrate this more fully than the sudden and unforeseen expansion of the fulmar across the North Atlantic world; this might even be called the most dramatic conquest of Britain since 1066.

Fulmars are the most characterful of seabirds. They seem to be constantly at play, especially in high winds, and appear to demonstrate an inexhaustible curiosity in humans, quietly approaching any boat at sea. Many photos from the kayak can be filed under 'Photobombed by Fulmar'. These birds are often written about as though they possess an unflappable mastery of the winds, but much of their personality comes from clumsiness. As they misjudge a gust and lose their poise, a huge webbed foot is thrust into the air; after a split second of feather-ruffling slapstick they'll be balanced on the breeze again.

The arrival of thousands of fulmars has transformed the experience of Havera from anything its crofters would have recognised. But fulmars are not the only change to the island's avifauna: the hilltop rigs where crops were grown have also been conquered. These thin strips of rich land were known and named in detail: responsibility for rigs called things like Da Peerie Wirlds, Da Hoolaplanks and Da Kokkiloori was circulated round crofters every season. Now, the space where kale, oats and bere (the traditional four-rowed barley of the Northern Isles) were grown is barely traceable beneath the same foliage that would be there had the island not been farmed. Ruling this feral domain is another victor in the struggles between species: the *bonxie*. Every rocky vantage, whether a corner of the old schoolhouse or a chunk detached from a long-flattened dyke, is now a watchtower for a bulky skua.

This two-century legacy of change is a profound demonstration of the entanglement of human and natural worlds. The Havera way of life was transformed – ended – not only by the ill-judged whims of distant landlords but by the movement of fish offshore, while many differences between Havera today and in the past are a result of the changing social lives – the histories – of birds. At the end of my Shetland journey, I spent a night in the shadow of the Havera windmill's twin, the fourth-century Broch of Mousa. This is a domineering monument to human belligerence, yet now its walls reverberate with the gentlest purring. Storm petrels nest in the cracks between stones, and as I settled down to sleep, hundreds of tiny stormies fluttered across my sight line. Masons making ready for war had unwittingly built the perfect hive for these sparrow-sized seafarers.

It is tragic to see abandoned places that were once filled with people, especially when (as in the case of Havera) their magnetic personalities – their pleasures and regrets – shine through recordings of their voices. Yet in a world where humans wage wars of conquest not just among themselves but on almost every species on the planet, it might be heartening to see the agency of animals reshaping realms to which humans are, more than ever, peripheral.

At about 3 a.m. on my night on Havera, heavy rain set in. As I paddled south, soaked to the bone and (for the first time in this journey) truly cold, the windmill remained on my horizon, only briefly hidden by the heaviest downpours. Both the red cliffs of the north and the fertile, low-lying isles were now left behind. A diverse geology, including complex whorls of multiple rocks, had taken over. Drongs were no longer the square-edged towers of St Magnus Bay, but rugged grey wedges like spittle-wreathed teeth. Few mishaps had occurred so far, but I'd now misjudged the battery level of my phone and was left unable to check tides and weather. I knew such accidents would happen quite often, but with the tidal challenges that lay ahead, this was not a time I would have chosen. Momentarily I thought there was a virtue to this failing: that it made my journey more 'authentic' (not a word I'd ever usually trust). But I quickly realised that every seafarer of the pre-digital age had resources to judge tides that I lacked. I had some familiarity with what to expect, built up over the last two weeks (I knew, for instance, to expect ebb tides in the afternoon) and I also had time: if I confronted hostile tides or weather I could, in theory, sit them out and consume the ample food and reading still stowed in my boat.

The sky began to clear as I passed St Ninian's Isle, linked to the mainland by its tombolo beach. This pretty strand of sand, lapped gently by the sea on either side, is perhaps the most famous landmark in Shetland. Like many of Shetland's spits and bars, it indicates an alarming reality: a drowned coastline that has not stopped sinking. Shetland may have sunk as much as nine metres in 5,000 years, in contrast to most Scottish coasts that follow a more usual post-glacial path of continuing ascent ('isostatic rebound' once freed from the weight of their Ice Age glaciers). The scale of this change, over so short a time, explains some of the extraordinary transitoriness of this coastline. This would be driven home even more strongly as I continued towards the final tidal barriers in my path.

The first challenge was Fitful Head. Yet again, the wind was low as I reached a point where any breeze would have spelled trouble, and yet again, this was enough to draw the sting from a possible threat. I bounced fast enough through the tidal overfalls at the Head to consider tackling the second challenge before dark. But Sumburgh Head was worth waiting for: it offered the possibility of whales and the chance to see one of Britain's most spectacular tidal runs, *da roost*, in action.

As the sun rose over another subdued sea – the fading swell preserving the memory of long-departed breezes – I launched. Porpoises, closer than any I'd seen so far, edged along the coast ahead. I soon passed the largest stretch of sand on Shetland. There is no hint when paddling past that this was once the thriving village of Broo. Landowners and tenants in the early eighteenth century began to note deterioration in the quality of their land. Soon, it was 'declared valueless'. By the mid-eighteenth century, the once-wealthy village had been obliterated beneath 'a small dusty kind of sand, which never possibly can rest, as the least puff of wind sets it all in motion, in the same manner as the drifting snows in winter'.[16] Caused by the climatic cooling of Europe's 'little ice age' (the 1690s were one of the coldest decades in the last millennium), this tragedy was the

most dramatic evidence I had seen so far of the scale and unpredict-
ability of transformation on these coasts.

Passing this eerie site, I soon found myself sandwiched between
tides, and forced to make split-second decisions about my route. At
the first asking, I got it wrong, choosing not to go round the island
of Horse Holm but to tackle the straits between the island and the
mainland. This felt like taking a bike without suspension down a
steep road of huge cobbles: the powerful tide was with me, but at
times I was afraid the huge overfalls might bury or even break the
kayak. My spare paddle was strapped to my deck in two pieces, but
it was clear, as overfalls wrenched at the one I held, that a second
without a paddle would be disastrous. I'm frequently surprised by
how short these infamous tidal runs tend to be: I thought I was at
the start of a long and harrowing ordeal when I found myself spat
out into placid water. After this, Sumburgh Head itself was straightforward. The sun appeared as I made my way out to sea, south of
the whole of Shetland, and for the first time in my journey I could
see people gazing down, bird-watching binoculars raised to assess
the small yellow form scraping across the sea. They were there in
numbers, I soon discovered, looking for orcas that had been sighted
the previous day.

I landed at the launch of the tiny Fair Isle ferry and crossed the
narrow neck of land behind Sumburgh Head. This took me to a
spot rich in historic remains, including Shetland's most dramatic
Viking tourist-draw: Jarlshof. The Norse traditions of this southern
tip of Shetland are nearly as rich as those of Unst. Even the tidal
stream I'd just swept through is rich in story. The *Orkneyinga Saga*
is a tale of competition between Norwegian earls for the coasts and
archipelagos of the North Atlantic; like most such sagas it is gripping and evocative but fiercely elitist, with barely a glimpse of
perspectives beyond those of its entitled male protagonists. In 1148,
the saga says, Earl Rognvald Kali Kolsson, ruler of the Northern

Isles, was travelling between Orkney and Norway. With breakers all around, he was forced to run ashore at the south of Shetland. Rognvald wandered local settlements, enjoying anonymity and frequently (as was his habit) breaking into verse. One day, he met a poor elderly man near Sumburgh Head. Learning that the man had been let down by a rowing companion, Rognvald (disguised in a white cowl) offered to help him fish. The two rowed out to Horse Holm making for the 'great stream of tide . . . and great whirling eddies' that I'd just swept along. As the old man fished, Rognvald's task should have been to skirt the tidal stream by rowing the boat against the eddies. Instead, he guided them deep into the turbulence where the fisherman began to draw up enormous fish, but soon cried out in terror 'Miserable was I and unlucky when I took dee today to row, for here I must die, and my fold are at home helpless and in poverty if I am lost.' Shouting 'Be cheerful man!', Rognvald rowed like a man possessed, eventually drawing them clear of the chaos and back to shore. Still incognito, he gave his share of the catch to the women and children preparing the fish on land, but then slipped on the rocks, provoking howls of mocking laughter. Rognvald muttered one of his verses, rendered here by the Orkney poet George Mackay Brown:

> You chorus of Sumburgh women, home with you now.
> Get back to your gutting and salting.
> Less of your mockery.
> Is this the way you treat a stranger?
>
> Think, if this beachcomber
> Hadn't strayed to this shore by chance
> Your dinner tables
> Would be a strewment of rattling whelkshells today.

Sumburgh women, never set staff or dog or hard word
On the tramp who stands at your door
It might be an angel,

Though here, with the Sumburgh querns grinding salt out there,
It was only a man in love with the sea,
Her beauty, her rage, her bounty,

One who knows that, all masks being off,
In heaven's eye
Earl is no different from a pool-dredging eater of winkles.

'They knew then', Mackay Brown writes,

> that the reckless benefactor was Earl Rognvald Kolson (nephew of
> St Magnus), one of the rarest most radiant characters in Norse history.
> A fragrance and brightness linger about all Rognvald's recorded
> doings and sayings, as if the long sun of northern summers had been
> kneaded into him.[17]

But *da roost* and Rognvald's antics are unusual: old stories tied to
Shetland landscapes are few, and documentation of Shetland's early
history is far sparser than that for other parts of Britain. From the
centuries when much of the landscape would have been named we
receive only the barest skeleton of events. The *Orkneyinga Saga* says
that Shetland was split from Orkney in the 1190s after a rebellion
of the 'Island-Beardies' (as Orcadians and Shetlanders apparently
called themselves). From then on, Shetland was largely left to its
own devices, although it changed hands (from Norway to Scotland)
in 1469. Only when the conditions of medieval Shetland began to
collapse, in the dire economic circumstances of the late sixteenth
century, are there detailed written records of what life here might

have been like: bitter complaints at the loss of order and well-being. By this point, Shetland was an exceptionally cosmopolitan place, the islands frequented by merchants from around Europe and the North Atlantic, so that Shetlanders often spoke some German and Dutch as well as their own Norn language.

The reason for the dearth of early Shetland stories is the eighteenth-century death of Norn. In literary terms, this loss was total: remarkably, no Norn literature survives except in second-hand fragment. Yet Shetland's dialect tradition is a worthy successor to the Norn heritage and a key impetus behind the wealth of current literature. This is undoubtedly the richest dialect in Britain and a constant presence in the experience of visitors (few, I imagine, leave Shetland without succumbing to the temptation to call small things 'peerie' or to replace 'th's with 'd's and 't's). It is among Shetland's greatest assets and the source of much of the archipelago's distinctiveness.

Once Norn died, dialect flourished. The nineteenth century has an awful reputation where dialect traditions are concerned: the bureaucratisation and centralisation of British life led many autocrats to think like Thomas Hardy's mayor of Casterbridge, who labelled dialect words 'terrible marks of the beast to the truly genteel'. Yet Shetland bucked the trend, forging – as always – a path all its own. By 1818, the crofters visited by Samuel Hibbert used words and grammatical constructions substantially the same as those employed by Shetlanders today (although, as the Shetland archivist Brian Smith puts it, 'naturally, the vocabulary is different, since we live in a society where dozens of words for small-boat equipment or seaweed, are unnecessary'). But the 1880s and 90s, during which Shetland crofters and fishers were freed by national legislation from the worst exploitation of landlords, marked a particular moment of growth. The first Shetland newspapers were founded in 1872 and 1885, and both specialised in dialect prose. They ran long serials such as 'Fireside

Cracks' (*Shetland Times*, 1897–1904) and 'Mansie's Rüd' (*Shetland News*, 1897–1914) which used island language to offer subtle observations of island life. 'My inteention', says the narrator of 'Mansie's Rüd', 'is no sae muckle ta wraet o' my warfare i' dis weary world, as to gie some sma' account o' da deleeberat observations o' an auld man, on men an' things in a kind o' general wye.' As this suggests, these columns were not inward-looking things, but helped form distinctive island perspectives on the world at large. In this newly prolific era, Shetlanders such as Laurence Williamson began collating and categorising dialect words and phrases, while others, such as the Faroese linguist Jakob Jakobsen, began to attempt to recover the old Norn language.

In this atmosphere of renewed self-confidence, some Shetlanders, such as the dialect poet William Porteous, began to use English to evangelise the islands to those mainlanders who, if they thought of Shetland at all, pictured a dreary scene. Poetry 'advertising' Shetland to the urban south embraced a flamboyant romantic aesthetic that wouldn't have been possible a century earlier. This marks, perhaps, the beginning of the idea that Shetland is the northernmost and fiercest expression of a frayed Atlantic edge worth celebrating. Ferocious storms, wind-whipped seas and bleak, unpeopled headlands could be romanticised, rather than being dismissed as incompatible with progress or politeness. In Porteous's descriptions of the 'strange exultant joy' to be found in confronting ocean weather, many touchstones of later evocations of the North Atlantic sublime can be found.

Two days after completing my descent of Shetland I found Porteous's verse in the local archives and was struck by his heroic efforts to make storms not just poetic, but an actual tourist draw. I felt that, having kayaked this far in improbably blissful calm, I still lacked a crucial aspect of the Shetland experience. But there was no need to have worried: five days after I rounded Sumburgh

Head a brief but grisly weather front was forecast. I'd already decided, thanks to Sally Huband's influence, that my last Shetland venture had to be to Foula. Given the risk of rough weather I chose not to paddle the crossing but booked my kayak onto the twice-a-week passenger ferry. At thirty feet long this is essentially just a diesel sixareen with a lid. It was perhaps surprising that only the youngest of the eight passengers was sick as, rocking and rolling, we meandered a slow course across the sea's contours. Minutes after disembarking in Foula's tiny harbour I launched the kayak and was paddling up the island's east coast. By the time I reached the north-east corner, I was in a sea terrorised by breakers three times my height, from which I could look east to the whole Atlantic coast I'd paddled down. That night I wandered through the Foula coast-line's meadows of ground-nesting seabirds, taking great care to find myself a spot to sleep away from any chicks (figure 2.7).

It was on the second of my three days on the island that I finally met the force that Porteous had pronounced Shetland's true ruler: the 'Storm King'. That morning I'd climbed the island's tallest cliffs and lounged, lodged among puffins, with my feet dangling 1,200 feet over the sea (figure 2.8). I sat gazing north while clouds gathered in the west and large, cold drops of water began to fall. Hundreds of screeching kittiwakes took to the air, disturbed by eddies in the wind, while spindrift skimmed the sea in all directions. The real arrival of the storm was preceded by minutes of strange, thick warmth. Then lightning flashed across the ocean, lending the swell fleeting new patterns of light and shadow, before rich, deep thunder reverberated through the rock. Puffins and fulmars joined the kittiwakes in panicked flight, and the booming cliffs themselves seemed to have come to life. The whole spectacle was as sublime and life-affirming as Porteous had, a century before, promised his urbane readers it would be:

And when the Storm King wakes from his sleep
  in the long, long winter night,
And, robed in his garment of silver spray, strides
  southward into the light –
At the sound of his voice ye shall see the waves
  race in for the land amain,
Then, broken and beaten on cliff and beach, fall
  back to the sea again.
Ye shall see the tide-race rise and rave, and rear
  on his thwarted path,
Till stack and skerry are ringed with the foam
  of the hungry ocean's wrath.
Ye shall watch with a strange, exultant joy, the
  winds and waters strive,
And your hearts shall sing with the rising gale,
  for the joy that you are alive.[18]

ORKNEY
(August)

Thermometer and barometer measure our seasons capriciously; the Orkney year should be seen rather as a stark drama of light and darkness.

George Mackay Brown

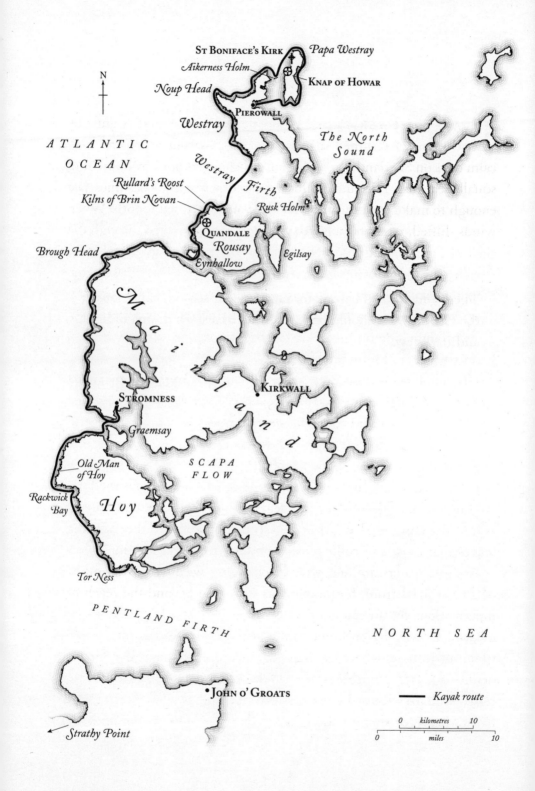

L ATE SUMMER BRINGS uncertainty. It was mid-August by the time I resumed my journey and, in contrast to the long calm days on Shetland, the Orcadian sea changed hour by hour. Sun, squall, sea mist, rain and rainbow passed across the coastlines fast enough to make each morning feel like a time-lapsed month. When winds shifted, water responded: weather was conducted through the shell of the boat and into me, dictating the experience of each new stretch of coast. Not just perception but emotion drifted with the moods of sky and sea. This ranged from the giddy joy of lurching along, propelled by a following swell, or the anxious focus when gusts brought side-on breakers, to serenity on flat seas that seemed perfectly safe and infinitely spacious.

Phases of transition from calm to chaos were often the most sublime, binding beauty and fear together. Not for nothing are the islands of Orkney said to evoke sleeping whales: they are peaceful mounds with awful potentials. As the month progressed I saw, felt, and, for the first time, photographed, parts of waves that seemed more the habitat of surfers than paddlers. These were not the long strafing breakers that come with heavy swell (I still had some leeway before the truly wild weather of autumn); they were the standing waves that twist and coil over any obstacle to a running tide.

Atlantic waters are deceptive in changing weather. In the midst of a tidal maelstrom, hospitable seas can seem beyond the reach of imagination; yet unseen gentleness might be just a few wave crests away. This was driven home to me at the north-west corner of the island of Rousay, where the sea's tidal features are named with the detail of a city suburb's streets. Here, emerging from a tide race called Rullard's Roost I hit a mesh of tide and swell so fierce that I had to head for shore: I thought my day was done within an hour of setting out. Yet five minutes later a more coastal line allowed me through: I could barely see evidence of conditions to cause concern. Much of successful kayaking is in the choice of routes between the shifting waves. As important on the water as arms or balance is a cool head through the roaring, swirling, chilling and grinding that batter the senses in a threatening sea.

Not just the weather, but also the landscapes were now defined by contrasts. On the first islands I passed, transitions from thundering cliffs to the placid undulation of cattle farms are sudden yet somehow seamless. No single landscape lasts more than a few hundred yards. On the most north-westerly island, Westray, the imposing, sixteenth-century edifice of Noltland Castle looms over a large modern farm; seen from the sea, the two occupy the same small space. Beside them, near the spot where surf meets sand, a sprawl of tyres and polythene marks a recently excavated sauna, built by the island's Bronze Age inhabitants.

This landscape looks at once spacious and cluttered. Centuries and functions, whether sacred, industrial, defensive or recreational, are pushed together at sparsely situated sites. Around them, in wide fields that are almost moorland, the earth is loaded with low-lying detritus of millennia. When I wandered ashore, I found myself watching each inch of ground for traces of the past until every broken plastic bucket or scrap of rope became an artefact. The sounds of breakers, cattle, lapwings, tractors and voices also took on that character: items in the soundscape felt as distinctive of this place as did objects in the landscape.

After making my way north by roads and ferries I had kayaked out from Pierowall, Westray's capital village. Its small grey buildings perch around a colourful little bay: tall yellow hawksbeard flowers and bronze kelp line an arc of golden sands and green sea. Pierowall was known to the Norse as Höfn and a row of pagan graves suggests it was a Viking-era market. When Rognvald Kali Kolsson stopped here there was a clashing of cultures: he met Irish monks whose hairstyles he mocked in verse. This pretty, ancient port was my place of departure but it wasn't northerly enough to be my true starting point.

I began by kayaking north-east. The small island of Papa Westray, known locally as Papay, thrusts a rugged and disruptive head north of Westray and into the Atlantic's flow. As I paddled into the mile-wide sound between islands I found myself grinning with pleasure to be back among the waves. I'd missed the ocean's noise, the tension in the arms as they pull a paddle through water, and, most of all, the sense of unrooting that rocking over waves creates. I kayaked carelessly, enlivened by cold splashes from the bow and paddle. Yet before I'd even really got started, I felt the lure of Papay's past.

This island proved to be the most improbable place I've visited. Its history emerges from waves and grasses in ways that feel surreal. Sometimes traces of the past are recent and mundane but still evocative of island life. My route reached the island at a pretty place where low cliffs are topped with a small, strangely situated structure that is blackened by burning. It stands on its isolated outcrop because this picturesque inlet faces directly into south-westerly wind and swell. For decades the vulnerability of this spot made it the ideal rubbish dump. Litter on the scale of cars and sofas would be thrown down the rocks and carried away by winter storms that were more muscular and reliable than any binmen. Local lobsters still dwell, perhaps, in the rusted boots and bonnets of Ford Cortinas.

As I rounded the island, the surprises became more venerable. I passed an enormous kelp store, a remnant of the decades round 1800 when Papay was a global centre of this major industry. Paddling past, and beneath one of the most spectacular chambered cairns in the world, I landed on a sandy beach beside a small and unassuming isthmus of stones and seaweed. I'd intended to wander up the cliffs and visit a monument to the extinction of an Atlantic seabird. But the spot I'd landed at was not what it seemed. At first, I thought I was hallucinating as I saw patterns in rocks where seaweed was strewn like tea leaves. But the more I stared, the clearer the geometry became: a cobbled platform took shape, then hints at a low stone wall. These sea-smoothed structures were centuries older than the era of kelp but, for now, the nature of their making remained a mystery.

I wandered up the cliffs to find the monument I'd stopped for. In 1813, 'King Auk' was the last great auk in Britain. These birds – penguin-sized relatives of the razorbill – were once prized for feathers, meat and eggs, but by the early nineteenth century the collection of stuffed birds had become a favourite pastime of Europe's elite. What could possibly cement a wealthy collector's status like the large, impressive corpse of 'the rarest bird in the world'? There are many discrepancies in narrations of the events of 1813, but it seems that 'local lads' had killed King Auk's mate by stoning the previous year. Now William Bullock, impresario and keeper of the Egyptian Hall in Piccadilly, had written to the lairds of Papay requesting the very last bird for his collection. The obliging lairds tasked six local men to row to the third cave along the north Papay crags. King Auk leapt from his perch into the sea and a marksman, Will Foulis, fired and fired again. But the auk was agile in the water. Eventually cornered, King Auk was bludgeoned to death with oars. The bloodied prize was soon in the hands of couriers to London where it became a feature of Bullock's ever more elaborate displays, to which another one-off, Napoleon's carriage from Waterloo, was later added.

The cairn I visited on the cliffs above King Auk's perch was put in place by local children. Concealed in the memorial, beneath a bright red sculpture of the royal bird, is a time capsule containing the message they wrote to the future:

We wish there was still a great auk to see. We hope that people won't have to build more cairns like this to remember things we see alive now. We humans gave a name to this bird, now only the name is left. If you who are reading this message are not human, remember us with kindness as we remember the great auk.[1]

The fate of King Auk marks Papay as a place of endings. But after I'd battled round the island's violent northern headland, I reached sites that spoke instead of beginnings. The most famous is the Knap of Howar. This is the earliest known constructed house in Europe. Built as a family farm around 5700 BC the land its occupants tilled and grazed has been eaten away by water until the Knap is nestled in reach of sea spray. Its concave walls and intricate cupboard-like enclaves are missing only soft furnishings and whale-rib rafters. Rabbits burrow all round. As they dig, they disinter refuse from ancient human meals: worn oyster shells, and great-auk bones, whose flesh was stripped millennia ago. Like so many sea-lapped sites, the Knap of Howar inspires conflicting responses. Thoughts are easily lured towards ideas of timelessness, yet everything about this site has been transformed: the quality of its earth and the nature of its foliage have been slowly altered by the creeping proximity of ocean. If timelessness exists anywhere on earth, it is not in sight of the sea.

Even the Knap of Howar is not the most immediate and affecting spot here. A little to the north, St Boniface's Kirk stands on the site of older holy places. Northerly gales flay earth from every inch of coast, changing topography by the week. Grasses and wildflowers cling to steep sandy soils where summer respite from storms provides the fleeting chance of growth before roots are ripped away and flung into autumn. It's easy to sit and stare into the ocean without comprehending the structures of rock and shell around you. From every inch of land the ocean takes, there appears a new facet of a large medieval settlement.[2] I'd glanced around layer upon layer of exposed walls and floors before I began to notice the refuse beneath them: thousands of shells of limpet, oyster and winkle clustered where they'd been littered after feasts. Storms here have disinterred whale vertebrae, from even grander feasting, and red quernstones for grinding grain, made of rock not native to the island. Remnants of the processing of pig iron and fish oil imply a community that worked the coast in sophisticated ways.

There's something evocative about the daily changes occurring at this unmarked, uncelebrated site. The configuration of buildings and shells seen on any visit is immediately taken by the ocean, never to be witnessed again nor recorded. It's impossible to categorise such places. Most of this island fits both poles of many binaries depending on the light you choose to see it in: human/wild, timeless/changing, productive/barren. Everything seems both out of place and perfectly positioned, and our frameworks for comprehending the coastal past feel entirely inadequate.

Unable to imagine what it must be like to live in a landscape so immediate but so inscrutable, I knew I needed help. Before setting out I'd contacted Papay's 'biographer', Jim Hewitson. Jim told me he and his wife Morag intended to travel no further than the Old Pier, 500 metres from their home, for the rest of the year: when I passed, he said, I'd find them at home or in a nearby field. In the early afternoon, I knocked on the Hewitsons' door and was led into an old schoolhouse. On one wall was a large map marked with Papay's historic place names. Elsewhere were images from the island's past including a painting of King Auk. This was pinned beside a memorial to a French kayaker who visited when paddling north. He'd planned his journey with his wife before her untimely death. Having undertaken the voyage alone, he disappeared, presumed drowned, before reaching Shetland; I didn't dare ask whether he and I were the only kayakers to have visited the Hewitsons.

I sat with Morag and Jim, consuming tales of island life along with tea and croissants. Then we wandered the coast. I was soon told, in no uncertain terms, that my desire to find explanations of the coast's mysteries was not an acceptable approach to the island. Life on Papay, they insisted, involves coming to terms with mystery, not seeing it as a problem to conquer. Jim and I revisited the strange cobbled structures of the beach I'd landed on. He told me that archaeologists call it a medieval fish farm, used by monks from a monastery that may or may not have existed when Papay might or might not have been the centre of an eighth-century bishopric founded by Iona monks (or someone else). Yet Jim and Morag's children have found antlers in this 'fish farm'; perhaps this was actually a spot for trapping deer whose movements would be impeded in the soft coastal ground. I was left wondering about the boundaries of 'mystery'. Without Morag and Jim's help it would have been impossible for me to write about the island at all. Before our conversations I had answers to bad questions; I was left with much better questions but no hope of answers. And I'd been given a reminder that archaeology is rarely about discovering or confirming facts, but more often a process of inventing the most plausible possible stories.

As we walked, pieces of Papay stone continually issued from Jim's pockets. One contained fossilised raindrops. Another was a Neolithic hammering tool. A third had been scratched at some inestimable date with a design that echoed the hills of Westray as seen from Papay. This was Jim's illustration of the power and persistence of island mysteries: it was probably – almost certainly – nothing, but it *might* be a rare piece of millennia-old representational art. I was reluctant to abandon a place with so many surfaces to scratch, and it was chastening to think I'd been tempted to kayak straight past. As a parting gift, Jim gave me an oyster shell. This was one of the many extravagantly ancient relics unearthed by the Knap of Howar's rabbits. Or else, perhaps, the oyster was alive and well until last year, when a black-backed gull had torn it from the seabed.

Orkney is often celebrated for the balance its people have sustained between the industries of land and sea. In comparison with Shetland, the more fertile earth shifts, slightly, the balance of subsistence onto the land and away from the ocean. To the islands' great bard, George Mackay Brown, Orcadians are 'fishermen with ploughs' although others suggest they're better described as crofters with lines and nets. The most celebrated Orkney historian, Willie Thomson, addressed the same theme with reference to the *Orkneyinga Saga*. He introduced Orcadian trade by evoking an ally of Earl Rognvald – Sweyn Asleifson – who is sometimes labelled 'the ultimate Viking'.[3] Like centuries of later Orkney folk, Thomson insists, Sweyn whiled away the year on his home island attending to agriculture; he only set out on ocean voyages in the interstices of the farming calendar. As I paddled from Papay to Westray, a tiny fishing boat motored back to Pierowall over flat blue sea; soon I saw a Westray woman herding cattle from a sea cliff to a gentle field a hundred yards inland. A huge bull bellowed its resistance. Never on this island was I out of earshot of either cattle or the chug of inshore vessels. Never did I find a coastal spot to sleep where I was certain cattle wouldn't appear around me.

Yet as I kayaked I became increasingly uncomfortable with the tradition of emphasising the contrasts and complements of land and sea. These coasts were thickly marked with remnants of industries at the margins. For centuries, every job at sea was matched by a dozen people working not the land, but the shore. If boats were constant protagonists in Shetland story and history, then the intertidal zone plays that role in Orkney: it runs through island literature in ways that are entirely unique. Memoir after memoir of Orkney life makes the shore a major character when boats are only incidental presences. A striking example was published by the poet Robert Rendall in 1963. This memoir, *Orkney Shore*, sold well on the islands, yet is almost unreadable today because of the knowledge it demands of Latin and dialect names for coastal species. Rendall compares his memoir to old-fashioned sugar candy held together by a central piece of string; his life, he says, is the uninteresting string, his depictions of the Orkney shore the delicious candy. A far more palatable, if emotionally challenging, memoir of coastal life, Amy Liptrot's account of recovery from addiction in *The Outrun*, brings the tradition of identifying Orkney with its shoreline up to date.

It's tempting to trace the origins of this theme back centuries. Whereas in most of Britain land ownership ended at the high-water mark, a different custom prevailed in Orkney: Udal Law, imported from Norway in the ninth century, extended kindred land rights to the lowest tidelines. Where in Scotland the intertidal zone was sea, in Orkney it was land. According to Ruth Little, director of a 2013 arts project called Sea Change, 'Orcadians are thresholders' whose access to the margins has defined their identities.[4] Even today, the conventions of Udal Law are sometimes successfully evoked against commercial threats to coastlines.

Many shoreline activities that families undertook related to fishing. Limpets were knocked off rocks for bait, nets were mended and lines prepared. Island women carried home the catch in heather creels before cleaning, splitting and drying fish. In a community where men were often offshore, Westray women performed many tasks that were elsewhere gendered male. 1920s photographs show women waist-deep in water hauling boats up Orkney beaches. They cut and carried peats, brought in hay and collected seabird eggs.

Groups of neighbours in this deeply social community would go down to the shore and collect seaweed, whelks and *spoots* (razor clams) or lay nets across the fields to dry.[5]

Many coastal tasks were distinct from both fishing and farming. My hope as I kayaked Westray's coasts was that I might teach myself to see the shores as resources. That leap of imagination into the perspectives of Orkney's past involved putting aside modern attitudes to eating puffins, bludgeoning seals, or spending the evening in a room lit and fragranced by blubber or fish-oil lamps.

As I reached Westray from Papay I passed a tiny skerry called Aikerness Holm (figure 3.2). This is nothing more than a flat pile of shattered flagstones in the ocean, yet a crudely built structure, like a misplaced garden shed, is perched upon it. I landed and looked round. Today, this would be unpleasant, cramped conditions for one; but here, in the nineteenth century, four or five men would spend their summer collecting seaweed with rakes and barrows, returning to Westray only at weekends. They'd burn heaps of seaweed, sending huge palls of blue-beige smoke floating to the island and obscuring sights and smells behind the infamous 'kelp reek'. The result of their burning was an alkali used in distant cities to make soap and glass.

Yet this tiny skerry is more famous for another major industry of the shoreline. On this spot, countless ships were wrecked. Later, in the archive, I'd listen to recordings of Westray folk describing aspects of island life.[6] The windfall of goods from Aikerness was prominent among their recollections: the most infectious guffaw to issue from an islander came from Tommy Rendall when asked the question 'Did any pilfering go on?' He told of errors made with things washed up from wrecks, such as the time when half the stoves of Westray were ruined because anthracite was mistaken for domestic coal. He told of customs men, whose task – to prevent the contents of wrecks from 'disappearing' – made them the most hated people in the islands (besides perhaps the lairds). Customs men were the butt of endless plots, tricks and jokes. Known locally as 'gadgers', these snooping officials are still recalled in Orkney descriptions of unruly children 'running round the hoose like a gadger'. But Westray's 'bounty of the sea' was in fact hard-earned. The people of the islands saved countless lives, rowing small boats out in all conditions to extricate crews from stranded vessels. Like much of island life, this was an improvised affair. Even in the early twentieth century the region's only sea rescue equipment was on Papay, because of the coincidence that the *City of Lincoln*, a ship large enough to carry such gear, had been wrecked there.

The first sea creatures I saw as I rounded Westray's northern headlands were seals. Whiskered snouts protruded from surf in almost every inlet. I'd soon discover Orcadian seals to be the friendliest and most playful I'd ever crossed paths with, but that's not because their relations with humans have been peaceable. Two days later I suddenly realised how many small structures I'd been paddling past were placed with sight lines to intertidal rocks where seals lounge. They were shooting stations (figure 3.3). Seal killing was once an enticing pursuit for Orcadian crofters: a single sealskin sometimes had the monetary value of a week's farm labour. And a seal served many other purposes, providing food, warmth, light from oil lamps and even protection for harvest machinery: anything vulnerable to rust was coated in seal fat for the winter. There is a cautionary tale for anyone tempted to see use of these marine-life fats and oils as 'traditional' or even 'barbaric' rather than 'modern': it was oil from north-east Atlantic basking sharks that lubricated moving parts in the Apollo moon missions.

It was not so much the import of cheap oils as new passions for wild animals that put an end to the seal trade. But recordings in the archive suggest the economic benefits of the seal to have changed rather than died out. One Westray resident, Alex Costie, recalled the end of seal hunting:

All the greenies, the likes of Greenpeace, were protesting so much
. . . that totally destroyed the markets, but I have discovered nowadays
how easy it is to get money for showing a tourist a seal that I am
now the most reformed seal hunter you would ever come across.

By the time I reached the end of my first day's travel I was at the
end of Westray's western peninsula, Noup Head. I climbed the
cliffs of this dramatic promontory and slept beneath an imposing
Victorian lighthouse. I was back among gannets. Shortly before I
came in to land, one eccentric bird approached my kayak and
clamped its beak around the bow before swimming alongside for
a while (figure 3.6). When I watched them from the cliffs, these
tardy birds – the last of the colony to leave for the ocean – were
exceptionally bad-tempered, like autumn wasps, protecting their
enclaves from each other with a noisy vigour I hadn't witnessed
before.

Next morning I awoke surrounded by half a dozen curlew and,
further away, a flock of lapwing. I steeled myself to the task of
imagining them as breakfast. Westray folk once used dried strips of
seal hide as rope for lowering islanders down from precisely the spot
I'd slept to snare birds on the cliff face. In the archive, I listened to
discussions of the subtle ethical considerations behind the collecting
of eggs and wildfowl. The first brood of lapwing eggs, Tommy
Rendall said, was always gathered in, but then lapwings were off
limits for the year: the second litter, being further into the summer,
was more likely to be raised successfully than the first. I was intrigued
to find that some of those interviewed had not entirely shaken off
old habits of seeing wildfowl as food:

The guillemots that came here, they still come here . . . you'll no
get any more here unless you build more cliffs because the cliffs
are full of them . . . It was always a great source of food for the

old folk you see. No expense, you didna have any vets' bills or anything . . . you know it is very dark-coloured flesh that's in them . . . sometimes they were just stewed but usually they were just boiled, you know boiled until the flesh fell off the bones, fried up with onions.

We used to eat eider ducks more than guillemots because there were more eider ducks in our area . . . and cormorants was better still, especially the brown ones, the juvenile ones. The meat in that is tender, better than any of the other birds I would say, apart from curlews . . . but nobody seems to eat that sort of thing nowadays. They are just dying of old age and going to waste.

As subsistence activities, these practices tend to evade the historical record. Never in British history has there been a market for the meat of young brown cormorants, however tender. The community activity of catching *spoots* on the biggest ebb tides of the year (for which children were even taken out of school) could produce a huge surplus of razor clams, but without refrigeration there was no potential for that to be exported either. Children might make a few pence from collecting whelks or catching coastal rabbits but that was the limit of such trades. These shoreline practices, unrecorded in tallies of import and export, are the great forgotten industries of Atlantic coasts. They were local, but far from peripheral because life itself depended on them.

The most marketable of traditional coastal pursuits is unsurprisingly the one that has survived. Every day I saw small creeling boats, most of which gave me a hearty wave as they motored through the tides. The potential to exchange lobsters for money means that not just fishermen or farmers have kept creels; for two centuries at least, almost everyone could supplement their income in this way. Many islanders recall collecting lobsters with particular pleasure: 'The smell o' the sea, a creel coming in with a lobster flopping, the tail banging about, it is a grand sound, a grand sight.' Some added that they didn't eat lobsters themselves ('well, perhaps just a small one'): these were seen not as food but money.

After Noup Head, Westray's dark cliffs alternate with gentle grassy slopes and long white sands. Farmed extensively but spectacularly un-intensively, each of these landscapes is stalked by sheep and large tawny cows. Between the modern farms on my skylines were many other abandoned buildings dating from a time of much more intensive usage of this landscape. Such ruins, with their sagging and crumpled flagstone roofs, attest to the slow exodus from the island. From over 2,000 residents in 1880, Westray had around 1,000 by 1940 and little over 500 by the turn of the millennium.

The 1930s were key to this process because two island industries collapsed. One was herring. From the mid-nineteenth century, fleets of drifters, like pods of orcas today, followed herring from the Western Isles to Orkney and Shetland. Their crews lived on ship and had limited contact with islanders, but Westray men took on the task of keeping fleets supplied with coal. The herring season saw the arrival of hundreds of women who gutted the fish. Unlike the men, they became fleetingly, precariously, integrated into Orkney life. As one islander, Jack Scott, recalled, 'suddenly, one beautiful day in summer 300 girls would appear . . . they were Gaelic-speakers and we didn't know what they were saying to us'. Scott went on to recount the pranks these women played on young island boys. The gutters were also associated with the arrival of exciting things: new Harris tweed suits for schoolboys and, for adults, exotic goods like cherry brandy and peppermint wine. After a summer of singing and accordion-playing the women were gone: 'it felt as flat as a flounder when they went'. Another island resident, Meg Fiddler, recalled the legacy they left behind in knitwear to last the year. Many photographs of these 1920s gutters show fashionably dressed women who look more like film stars than modern prejudice against the smell of herring might lead people to assume. A slump in herring numbers signalled the industry's demise. In 1939, the buildings used by gutters and sales agents were commandeered for the war effort and, for Orcadians at least, the industry was dead.

Kelp was another rich trade that hit hard times: this was an export entirely dependent on the whims of distant industries. At the peaks of a kelp boom whole families helped build huge piles for burning. Westray and Papay were as alive with the smoke and fire of industry as Manchester or Coalbrookdale. Orkney was unique in making large local fortunes from kelp. Elsewhere, aristocratic lairds considered trade unseemly so rented the shore to incoming kelp crews. But Orkney's merchant lairds pursued the trade with their resident workforce. These landowners could manipulate labour with ease because many Orcadian tenants paid rent in labour rather than money or goods. Their lives involved being constantly *on ca'*, moving at the laird's command between tasks of land, coast and sea. This is one reason why remnants of the kelp trade litter Orcadian shores. Most such ruins are from the first kelp boom after 1750; the end of this glut, in the 1830s, sowed some seeds of Westray's downward demographics. But other structures belong to a second boom when demand for iodine between 1880 and 1930 resurrected the trade.

Few people undertook the hard, unpleasant work of making kelp unless they were forced to, but the experience of compulsion varied according to the character of the lairds. The Balfours who owned much of Westray were not, it seems, especially unkind: 'You never got good lairds', Tommy Rendall noted, 'but the ones we got here were maybe the least bad ones.' Across the narrow Sound of Papay, the Traills were fierce autocrats who worked their tenants hard. Countless grisly stories are still told of them. There's the tale of a cruel Traill who was thought to have died until knocking was heard from the coffin at his burial; without a word exchanged, the only people close enough to hear – the crofters forced to carry the box – lowered him into the ground anyway. Another Traill was supposedly so corrupt that plants refused to grow on his grave in the Papay cemetery.

These stories were just a few in an array I heard while on these islands. Storytelling is, in fact, among the biggest and most beguiling industries of this shoreline. Few forces generate the serendipity of story as prolifically as the capricious and connecting sea. Even my boat provoked tales. When I arrived on the island, a Westray man looked my kayak up and down and told me that this was the first place in Britain to see such a thing. He dated this improbable event to an even more improbable date: 1682. Foolishly, I mistook this for an odd joke and failed to press him with questions. Yet the idea stuck with me enough to look for it in the small archive on the island. I found that the story of 'Finn-men' arriving by kayak in the 1680s was a venerable one. In a book of 1939 Iain Anderson wrote:

> Their appearance was, of course, almost unaccountable to these islanders, who recorded that their boats appeared to be made of fish skins, and so built that they could never sink. I think it may be accepted that these strange visitants must have been Eskimos who had been blown to sea when fishing off their own coasts. What seems to be most remarkable is that the Finn-men when seen in the vicinity of this island were still alive, and that when the islanders attempted to catch one of them, he escaped with ease owing to the speed of his kayak.[7]

These kayakers, if they were truly here, were as likely to have been Sami people from Finland as Inuits brought by the North Atlantic Drift. But by the time I reached the archive I'd come to terms with the idea that a historian's critical faculties needed to be used for purposes other than sifting truth from falsehood: deciphering the meaning of Westray stories was a subtler affair altogether. I'd met a dark-haired man who claimed to be descended from 'dons' of the Armada stranded here in the sixteenth century. I'd heard tales of Westray 'whale shepherds' herding pods of 300 cetaceans into local bays to take their teeth, and I'd heard the strange story of Archie Angel. This young boy had been discovered on the Westray shore after the wrecking of a Russian ship. He was named when the name plate of the ship, *The Archangel*, was discovered in the sea. Archie was integrated into Westray society so that generations of islanders had the surname 'Angel'. A host of things make this story unlikely (how did the islanders read a Cyrillic name plate?), but they are all beside the point. In a place where people washed ashore have so often played roles in the community, and where many houses have timber from wrecked ships built into their structures, sea stories shape island identity: the Just So stories of Westray life. In these tales, facts that can neither be verified nor falsified, yet have a certain pedigree, are the most powerful ingredients of all. The way in which history shapes Orcadian identities through stories and everyday artefacts feels somehow more immediate and pervasive than in anywhere else I've travelled.

Every month of my journey introduced new aspects of the Atlantic. The most immediate difference between kayaking Shetland and Orkney was the sea crossings. The main island chain of Shetland is packed tightly together. Although deep and treacherous, the drowned valleys that bisect the ancient mountains are narrow. In most places, crossing as the tide turns means there's little to worry about: each tricky stretch can be traversed in the time it takes the tide to reassert itself. Not so in Orkney. Although the islands are smaller, the distances between them are greater and the behaviour of the sea is more complex as it fills and vacates the inter-island gaps. Whether in ebb or flood, tidal flows coil back upon themselves. These eddies draw beguiling patterns on the water. Shimmering silver discs like pools of mercury pass through zones of dark ruffles. Bubbles, as if from the snout of a giant sea beast, rise where eddies meet. Veins, ridges, crests and watery fins drift slowly across the surface. The forces of swell, chop, tide and eddy sometimes work in concert, amassing as great heaps of sea. At other times they work in counterpoint, becoming complex cross-rhythms in an oceanic fugue.

Centuries of Orkney seamen have each spent years learning the major 'tide sets' of their area because – contrary to popular belief – tides aren't regular or predictable. As one seasoned Orcadian, Gary Miller, puts it:

> You get a tidal prediction book but that's all it is . . . they could be a lot stronger, if you've got a higher or low air pressure it can alter the tides, or the temperature of the water or the weather or if it's been windy . . . there's that many variables.

Learning tides meant learning which movements arrive early if a headwind is blowing, and in which regions water might run against prevailing flows. Local seafarers can explain everything of the tides around them. But for a kayaker passing through, these performances are yet more Orkney mysteries: tidal events defy logic like the acts of some inscrutable and wayward will. It's hard to believe this pulsing, breathing sea isn't alive. It feels far more superstitious to think that the interplay of cosmic orbs is weaving localised motions that – in this very moment – force your bow to buck and twitch.

Leaving Westray to cross to the island of Rousay was my first tidal challenge. From Westray's western cliffs I headed east between the headland at Langskail and the rocks called Skea Skerries. From here I could see the skerry of Rusk Holm, where the 'holmie' sheep graze seaweed, and a nineteenth-century tower was built for them to climb to safety when seas submerge their 'pasture'. I continued until almost at the south-easterly extremity of the island, then turned my bow south into the firth and steeled myself to paddle hard for Rousay's north-east headland. The golden sun was low, casting dazzling light across close and foamy ridges of sea, and with wind entering the firth from the east, a messy chop moved against swell that came in from the west. Small waves crossed large waves, merging and birthing pyramidal wavelets. These conditions conspired to make tidal movement impossible to read but easy to feel: the kayak's bow and stern took on minds of their own and my energy was spent less in moving forwards than in keeping my course. But the crossing was quicker than I'd feared (just a taste of what was to come). The particular local threat was that reaching Rousay offered no respite, because this island is the fixed point in a vortex of tides. Its headlands are sticks thrust between the spokes of a turning tidal wheel. It was here, after landing for the night, that I was forced to retry the tricky headland at Rullard's Roost.

The Rousay coast is famous among historians. Known as 'the Egypt of the North', its number of ruins is matched only by the volume of stories that arise from them. The sounds to Rousay's east and south are its relic-lined Nile. The small isles in the tidal river are as historic as Elephantine or the cataracts south of Aswan. Prominent among them are Orkney's two holy islands, Egilsay and Eynhallow. There is no landscape in Britain, besides perhaps the Wiltshire henges, which matches this few square miles for historic depth and diversity.

After a tidal battle at Rousay's north-western corner I kayaked through freaks of deep time. Wherever the joins in the Devonian sandstone are weak, caves, arches and gloups have formed. The grey, cream and ochre bands of rock – perfectly horizontal – are deeply pitted, leaving narrow pillars of stone, striped like Neopolitan ice cream, to support the cliff face. An airy space, the galleries of a dark drowned Parthenon, stands behind. The gaps between pillars have old, dramatic names like the Kilns of Brin Novan. The largest such 'kiln' is thirty metres deep by fifteen wide: within it, swell churns until it bubbles as if boiling. This movement threatened to pull me in as I hung at its mouth to marvel at the fracturing, scarring and sagging that the sea inflicted. These geological creations felt like the imaginary future ruins of a civilisation lost to the rising seas of the Anthropocene.

But the most remarkable features of this landscape belong to the shorter span of time between prehistory and the Victorians. I soon had vistas across the parish of Quandale, where old abandoned townships are sandwiched between the Atlantic and the hilly moorland called the Brae of Moan. It was only here that the true tragedy of the survival of Rousay's archaeological heritage struck me. This landscape survives in historic forms because it was once emptied by force. The region I was now kayaking was the only part of Orkney subjected to large-scale clearances in the nineteenth century. It was never transformed by subsequent development, because it was rendered barren by the design of lairds. After its emptying, this became a spacious sheep run offering a few pounds a year for little effort and less responsibility. What remains is a tapestry of overgrown dykes, runrig and small kale yards from which remnants of ruined crofts and silhouettes of prehistoric earthworks loom.

This was once an ancestral landscape and a world formed round burial mounds: it was a sacred place. In the Bronze Age, barrows were built to be visible from dwellings so that the dead continued to occupy the world of the living. Burned mounds also punctuate the hillsides. These are large piles of stones that were heated to boil water (although, since food remains are not found with them, their ultimate function remains obscure). The placement and scale of those in Quandale shows that, like the burial mounds, they were part of the social landscape, acting perhaps to display wealth or status. And the remains of millennia are intertwined. Views from the doorways of eighteenth-century farmsteads were dominated by some of the biggest burned mounds in the world. On Eynhallow, crofts were even built into the ruins of a twelfth-century chapel. This clustering of buildings – sacred, ceremonial, domestic – is not just due to centuries of similar uses of land and sea, but also to active relationships with the past among later inhabitants: folk beliefs, cosmologies and identities were shaped by life in a Norse and Neolithic landscape.[8] Quandale never ceased to be sacred.

To occupants of Northern Isles farmsteads, relics and monuments belonged as much to the present as the past. Ancient things were recycled into new buildings in ways that were ritualistic. Prehistoric axes were deposited in chimney stacks to protect houses from lightning strikes; Pictish symbol stones were built into thresholds and fireplaces, as were prehistoric cup-marks and spirals. Seasonal tasks, such as cutting peat, planting kale or bringing animals in for winter involved wandering different routes through the historic landscape. This resulted in a seasonally shifting geography of life that is sometimes called the 'taskscape'.[9] The farming cycle dictated which ruins were encountered day to day, encouraging seasonal repertoires of stories about the origins and meanings of ancient features. Communal memory was long, and stories that could explain how landscapes reached their present state were particularly resilient; since the end of the nineteenth century, Orkney has had an unrivalled number of folklorists, from Ernest Marwick to Tom Muir, who piece this scattered island memory back together.

Over time, townships expanded and the ancient features outside boundary dykes were drawn into the familiar and domestic world. These changes were never without meaning. Mounds, in particular, weren't neutral features in the landscape: they were sites at which the world of humans intersected with that of supernatural creatures called *trows* and *hogboon*.[10] The biggest mound in Quandale, the

Knowe of Dale, figures in Orkney tales of human abduction by the *trows*. Throughout the British Isles, uncanny associations caused farm boundaries to be sharply diverted round prominent mounds. This is why, as I kayaked past, I was surprised to see farms and mounds in close conjunction: at least two Quandale farms were built with barrows at their entrances. One such farm is called Knapknowes, which in Old Norse means 'Mound mound'. Even the Knowe of Dale is situated prominently within the township itself. The decision to do these things would not have been taken lightly: the barrows of Quandale, it seems, were given different meanings than those elsewhere. Because the long chain of Quandale memory was severed by the eviction of its people in 1848, even the most accomplished folklorists cannot reveal the nature of that difference.

Estate maps of this region around 1850 depict the area of the township, relabelled 'Quandale Park', as empty, showing nothing of the recently abandoned crofts or ancient sites. This was indicative of successive landowners' attitude to the land: they took great pains to present it as a resource, not as a place with history, traditions and stories. When Quandale did, eventually, reappear as a focus of their interest it was as a playground for indulging antiquarian fads. Many Rousay mounds have indentations in the top where Edwardian landowners and wealthy tourists indulged their passion for relic-hunting. Yet this early excavation arrived later in Quandale than elsewhere: for two generations the land remained too contested for lairds to be willing to show interest in tradition. Walter Grant, the first new laird of the twentieth century, was one of a pair labelled 'the broch boys' for their efforts to recover Bronze Age monuments. To Grant, however, a mound was an object to be described in isolation: the subtlety of the sacred landscape, including the complex interplay of its remains from different eras, evaded him and those who followed. Only recently, in the hands of innovative archaeologists such as Antonia Thomas and Dan Lee has the full complexity of this island's past begun to be understood. Today, the sacredness of Quandale is in its emptiness. The holiest sites, perhaps, are the nineteenth-century ruins: monuments to the victims of the lairds.

From Rousay I crossed to Eynhallow and wandered its short, circular coastline. The twelfth-century chapel here is another exceptionally atmospheric ruin. It is an intricate but crudely built holy place that looks out upon the fiercest tides. I climbed the chapel walls to view the terrifying overfalls that cut the island off from both sides. Folklore holds that Eynhallow was once enchanted and inaccessible to humans: its occupants were magical Finn-men. They called it Hildaland and were banished, by salt and the sign of the cross, only at the time the church was built. After its sanctification, Eynhallow earth was said to repel even mice and rats so that a bag of the island's ground became a valuable commodity.

It's still easy to believe that the fractious white water of the sounds is an enchantment made to hold Eynhallow at arm's length from the human world. Fulmars and seals take advantage of the safety provided by the tides, patrolling every section of the shore like guardians. I hung around on a patch of still water in the island's sheltered eastern bay, waiting for the tide to turn, as a group of tiny harbour seals swam repeatedly around and beneath. Glistening round heads came close enough for long whiskers to brush the kayak, their gentle breath audible as they surfaced (figure 3.5). With a warming sun and clear green water rippling over shell sand, this was the perfect tonic to trials by tide: my last moment within the sphere of the enchantment.

The contrast as I rounded the north-west mainland a little later couldn't have been greater. The wind peaked at sunset, bringing untidy seas (figure 3.4) and forcing a crunching landing into a black, rocky shore. Swings in the weather didn't let up until I left the islands. The next day, which took me down the western edge of the mainland, began in froth and sea spray. A scarred bull dolphin sped below, warning me away from its pod passing further offshore. Then I passed multitudes of rocky protuberances and crevices; these were bleak but colourful on an afternoon that truly was 'a stark drama of light and darkness' (figure 3.7). The day culminated in calm views over the final leg of the journey, including the most famous sandstone stack in Britain: the Old Man of Hoy. I turned inshore at dusk, with Hoy's red cliffs reflected in the sea to starboard and to port the twinkling lights of Stromness. But before I could explore the book-shop, cafés and arts centre of the first substantial town on my kayak down the coasts, I had to face one of my biggest challenges: the journey between the unrelenting cliffs and tides of Hoy.

Hoy took me two attempts. On my first effort to breach this most treacherous stretch of waters, I tried to take the sting from the crossing by spending the night on Graemsay. I slept by a disused jetty on this small island in the centre of the sound, with Graemsay's two lighthouses in sight and views across to the orange tinge of Stromness street lamps on the low blue clouds. Despite my precautions, I hit enormous overfalls at Hoy's north-western corner and was forced back. Even the inglorious retreat to Stromness cost me all the energy, strength and composure I had. I couldn't help but berate myself. On a sunny day with a gentle breeze my planning had been spectacularly poor. On the two days that followed, winds raged. I waited them out in town, taking the chance to talk with experts in aspects of Orkney and to plan Hoy properly.

On the third day, I set out in low wind but thick fog and rain. Visibility was poor and the waters starkly contrasting. In most regions of the sea, glossy and slowly rolling waves were gentle and rhythmic; but crashing cross-rhythms resulted wherever rock challenged the will of the water. Listening was my chief tool of navigation through the mist, and I was soon immersed in the patterns that lapped the edges of my boat. By the time I reached Hoy's cliffs, I was surrounded by the boom of breaking waves, listening hard for corridors of silence through the noise.

I took one break during the day, in the only major breach in Hoy's western cliffs. This wide bay is 'the Orkney riviera' of Rackwick, a collection of eighteen crofts and a schoolhouse where generations of Stromness folk once took summer holidays. My landing was through surf, and the launch back out from the beach of boulders was challenging. Rhythm was everything: processions of breakers a few feet tall alternated with short spells of waves at least twice their height. The troughs between waves revealed rocks beneath the water: obstacles that would make it hard to meet the waves head-on. If a spell of large waves and deep troughs appeared when I launched, my situation would be perilous: I sat listening for almost an hour, trying to find patterns and make predictions.

This day-long need to listen intently might, elsewhere, have been a chore. But here it felt like an opportunity. Hoy's waves have perhaps the most famous patterns and rhythms in the north-east Atlantic; hearing their refraction through art had been, twenty years earlier, the start of my engagement with the islands. My dad was a violinist in the BBC Philharmonic Orchestra in Manchester, where the resident composer was Peter Maxwell Davies, known universally as Max. In my teens, Max had given me lessons in playing his Orkney-inspired music. Until he passed away shortly before my journey began, Max was one of the most significant composers of his era, and Hoy was pivotal to his career. Before he found Orkney, he was the *enfant terrible* of British music, scandalising metropolitan audiences. In 1971 Max moved to the most remote croft on Hoy, high on the cliffs above Rackwick Bay and a few hundred yards from where I'd landed today. From here he took a leading role in Orcadian life, founding the St Magnus music festival, developing a new Orcadian musical style and evangelising Orkney to global audiences. Each year, my dad's orchestra would travel to the St Magnus festival and Max would come to Manchester to conduct works inspired by these waters.

To say that these works were 'inspired by' the sea doesn't do justice to the ways in which the Atlantic soaks them through. What separated Max from his peers was his sensitivity to the soundscapes of his environment. He arrived on Hoy in flight from the aural clutter of the city, but his new life wasn't defined by an absence of noise. It took place in a soundscape that proved far more provocative than he could have imagined. Surf rolls in on both the right and left of the home he chose: the sea and wind here are constant and inescapable but infinitely various. Gradually, Max realised their potential not as surface details, but as the generative force of his art. The sea became his answer to the puzzle that faced all writers for orchestra in the twentieth century: how to compose in ways that resonate with audiences without retreading the classical patterns of previous centuries. Most new methods, such as the twelve-tone music of Arnold Schönberg, proved to be rewarding games for composers but too abstract and cerebral for their patterns to be evident to listeners. Max's epiphany on Hoy was that the movements of the sea contained a balance between regularity and randomness that was ideal for generating music. Wave forms and sea rhythms were familiar enough to root listeners in experience, yet complex and alien enough to cause shock, wonder and revelation.

Max's seascapes are far from gentle and reflective, conjuring instead the roaring ocean thrashing at the Rackwick cliffs. Every cross-rhythm and complexity at the intersections of seas is intensified rather than simplified. These works are reminders that the ocean is only occasionally a soothing, pleasant place, and they capture the persistence of its presence in the Orcadian soundscape. When I listen to them now I'm drawn back into coastal nights in the sleeping bag, when the sea roared far louder than traffic outside any urban window. What the day sounds and feels like on Hoy is defined by the mood of the ocean. When the wind is up and the swell rolls in, the water's power is impossible to ignore, even from inside a house or the island interior's moorland. Ocean might be the background to all Orcadian life but it is the foreground to the sensory experience of Hoy. Tim Robinson, assiduous chronicler of the shores of Connemara, has perhaps done more than any other author to capture the noise of Atlantic coastlines in which, he states, 'only the most analytic listening can separate its elements'. Robinson's writing on this theme draws on the instincts developed during the training as a mathematician and career as an artist which predate his immersion in ocean sound-scapes. These challenging sounds,

are produced by fluid generalities impacting on intricate concrete particulars. As the wave or wind breaks around a headland, a wood, a boulder, a tree trunk, a pebble, a twig, a wisp of seaweed or a microscopic hair on a leaf, the streamlines are split apart, flung against each other, compressed in narrows, knotted in vortices. The ear constructs another wholeness out of the reiterated fragmentation of pitches, and it can be terrible, this wide range of frequencies coalescing into something approaching the auditory chaos and incoherence that sound engineers call white noise. A zero of information content.[11]

But no prose could pick out the order in the apparent incoherence of water-noise with the precision and richness of slowly unfolding symphonic music. Max was an obsessive observer, with the pattern-finding skills of a mathematician (there were far more books about maths than music in his home) and he studied these waves intensively over decades. As well as forming his music from the patterns of waves and of seabirds spiralling into the sky, Max filled his music with artefacts of the Orkney soundscape. Curlews, gulls and features of the weather suddenly emerge from the orchestral background. And fused with themes from the natural world are eight millennia of Orkney poetry and story. His subject matter included the runic inscriptions at Maes Howe, the tale of St Magnus, told as the story of a pacifist Viking, and the 1980s battle against the exploitation of Orkney uranium. His close collaboration with George Mackay Brown became the warm, social counterpoint to the cold inhuman ocean in an output of over a hundred musical seascapes.[12] And, like the Rousay crofters, he reworked millennia of Orkney history for present purposes.

Max isn't alone among Britain's leading composers in being drawn to Orkney: there's something about these complex waters that seems uniquely inspirational for music. Once I reached the south of Hoy, the mist cleared into a rich, bright evening. To the south-west, I could see the Scottish mainland. A dazzling white shard on the horizon was the lighthouse at Strathy Point. Its old engine room is now the home of the composer Errollyn Wallen. Born in Belize, overlooking the islands in the Caribbean Sea, Wallen now lives at the other end of the Gulf Stream, on the Scottish coast overlooking Orkney. In a song cycle, *Black Apostrophe*, inspired by Scotland's Caribbean connections, she set the seafaring poetry of a Bahamian-born sculptor, Ian Hamilton Finlay, whose life had also followed the Stream. Finlay had briefly been a labourer on Rousay: an instinctual link to his maritime childhood pulled him north from Edinburgh in the 1950s.[13] To both Finlay and Wallen Scotland is a sea zone and Orkney distils its archipelagic state. Given the power of water to this verse and music perhaps 'aquapelagic' is a better term: these island assemblages are defined by what lies between them. When Finlay left Orkney, he tried to take the waters with him.[14] He named the windblown ash tree by his inland window 'Mare Nostrum' (Our Sea), noting 'Tree and Sea are the same in Sound.' He referred to Nassau as his birthplace but Rousay his 'birthplace as a poet'. The rest of his life was lived in lowlands, but the boats and tides of Nassau and Rousay infiltrated all he did.

Wallen set two Finlay poems in *Black Apostrophe*. One was 'Fishing from the back of Rousay' which begins a thousand miles away where rollers, loud, relentless and unpredictable, 'Originate, and roll – like rolling graves – / Towards these umber cliffs'. They crash into land among weed-robed rocks, 'like sloppy ice (but slippier)', where limpets are the only frictive aid against a sideways slide into waves 'that rise and swell / And swell some more and swell: you cannot tell / If this will fall (Boom) where the last one fell / Or (Crash) on your own head'.

Like Finlay's, Wallen's sea joins land masses. It's a conduit between elements of her aquapelagic experience: 'I often ask myself "how did I get here?",' she writes, 'and I always answer "the sea".' The music of *Black Apostrophe* is united by the sense of a rolling swell, over which evocations shift between Belizian 'lush tropics' and Orcadian 'bleak majesty', the latter conveyed as much by complex harmonies as by rhythm: 'it was the sense of crossing the water for a world "out there" that I . . . wanted to capture'. The sea is a site of possibility and longing, evoking her parents' desire, in Belize, to cross the Atlantic, and her own wish, in Scotland, to feel the connection between archipelagos.

In listening to the same sea, Max and Wallen sensed different histories. Max heard centuries of explosive dissipation on the Orkney shore: his waves are at their moment of fulfilment, when switches are flicked between violence and silence. Wallen heard instead the sea's slow accumulation: its transmutation in the long course of travel bringing countless echoes of elsewhere. The Barbadian author Kamau Brathwaite calls the motions of connecting waters 'tidalectics': 'tossings, across and between seas, of people, things, processes and affects'.[15] And it's worth recalling that even the puffin – now emblematic of the north-east Atlantic – is a bird of the Pacific that, 50,000 years ago, crossed the cold waters that once parted North from South America before the Caribbean basin formed. Only with the birth of the Gulf of Mexico and Caribbean Sea did the Atlantic puffin become distinctively Atlantic. Only then did the warm currents gather from which the Gulf Stream now surges: the gentle climate of Orkney is made by the shores of Belize.

The work of Finlay and Wallen is also a reminder that it's misleading to think of Britain as a nation that had an empire or acquired an empire; Britain was born from an unequal union in 1707, when two colony-owning states – England and Scotland – were conjoined. From its beginnings, Britain *was* an empire and the sea was its medium. Money made from west-Atlantic slave plantations was used by British landowners to impose authority on Orcadian populations, and wealth made by those landlords from Orcadian kelp ran the machinery of slavery. Many families who owned Orkney land were connected as closely to India as the Caribbean. Indeed, the South Atlantic sea route, round the Cape of Good Hope, took ships to regions that had more places named after the infamous Traills than Papa Westray or Rousay where the family long held sway: Traill's Pass, for instance, leads not through Orkney hills but above the Pindari Glacier in the central Himalayas. The elites of Victorian Edinburgh and Glasgow understood the specific textures of places in the East Indies and West Indies better than the diversity of Scotland's seaboard and knew those places to be far more central to British fortunes than anywhere north of Scotland's central belt. It's no coincidence, then, that when Robert Rendall compared Orkney shores to the sugar candy of his childhood he unconsciously used a Caribbean staple to stand for the island nature of his home; in the sound of Rendall's crunching candy, as much as in the music of Wallen, there echo a thousand stories of an ocean-wide, aquapelagic, world.

THE WESTERN ISLES
(September/October)

ATLANTIC
OCEAN

N

Butt of Lewis
DÙN ÈISTEAN

Ness

Little Bernera
West Loch Ròg
Pabbay Mor

STORNOWAY
CALLANISH
STONES

Lewis

The Minch

Scarp

An Clisham

Taransay
TARBERT

Harris

LEVERBURGH

Pabbay

Vallay
Sound of Harris

North Uist
The Little
Minch

Monach Isles

Benbecula

Skye

South Uist

Eriskay

Barra

Vatersay
CASTLEBAY

Pabbay
Sandray

Mingulay

Barra Head

——— Kayak route

0   kilometres   25

0   miles   25

F ROM THE SIXTH century to the twenty-first the long chain of Western Isles, which stretches 130 miles from the Butt of Lewis in the north to Barra Head in the south, has been pivotal to the formation of North Atlantic cultures. These islands are marked by their early-medieval role as sites where 'thalassocracies' – the sea superpowers of Norway and Ireland – competed for control. Lewis seems so much like a Gaelic-speaking twin to Nordic Orkney that my leap from Scotland's east to west felt, but for the language spoken, like a short exercise in island hopping. Catholic Barra, however, is far more like an Irish island than anything that might be encountered in the north. The cultural difference between Lewis and Barra thus exceeds anything the distance would imply. But an outsider's experience of travelling these diverse islands today is defined by language. As the only great expanse of land where Scottish Gaelic is the medium of life for thousands, this is the primary site in which the tongue's future is defined. The isles, in all their contrasts, are thus united by a rich sense of history and a vigorous commitment to community and culture. This vibrancy has much to teach historians. The Western Isles in 1970 – hog-tied by national policies that paid no heed to local variation – were not the thriving place they've become. The last half-century has seen dramatic rejuvenation that makes this region a model for how peculiarities of place can be assets for modern, global life.

But the processes that shaped these cultures reach back beyond historic travels of the first Irish monks, and there's no way to read the islands' pasts without grasping the geographies that shaped the ebb and flow of local fortunes. This western geohistory is as different from the young, mutating archipelagos of Orkney and Shetland as it's possible to be. Places such as Barra give the impression of impossible permanence: they're entirely ancient bedrock that has lain, unyielding, since before the birth of the Atlantic. Large expanses feature few obvious glacial scars: these rocks seem barely to have registered a mile-high pile of ice grind over them. Seventy million years ago, volcanic chaos accompanied the opening of the Atlantic. From the traumas that separated Scotland from Labrador the laval fangs of the Inner Hebrides were born: the mountains of Skye, Rum and Mull are young rock cascades suspended in motionless pouring. But even ructions on this scale were too superficial to cause much change in the old, hard gneiss of Barra. The Outer Hebrides look on a geological chart like a timeless, providential flood wall, built to take the oceanic savagery that would otherwise shred soft tissues of the mainland.

The human consequences of this resistant rock are extensive. Focused on their stand-off with ocean storms, the stone has little left to give the land. The sea corrodes all exposed iron, breaking cars to atoms, but the rock lies incorruptible: nutrients leached away aren't replaced by slow mineral breakdown. For all it gives the earth, this gneiss might as well be the concrete of a car park. Such stubbornness is a problem for those who wish to make a living from moorland crofts, but it also stymies those who try to read the land. Since few peoples proved able to inscribe this stone, the cup-marks, whorls, crosses and scripts etched along the Atlantic seaboard are absent even from huge sacred monuments like the Callanish stone circle. A standard tool of historic interpretation is missing. So an oral culture, preserving stories through centuries, plays a special role in efforts to interpret the landscape: the skills required of historians, and the attitudes taken to evidence, are different from those employed else-where.

Paradoxically, this bleak landscape defines the islands' sheltered east better than the wilder west, where a strange quirk of the Atlantic transforms the nature of the coast. At sites such as Broo in Shetland, I'd seen arid sand that blew in from the ocean to smother fertile earth. But in the Western Isles the relationship of land with sea is reversed, and nowhere is the providence of the ocean more immediate: Cuan Siar (the Gaelic Atlantic) implies very different things from Nort Atlantik in Shetlandic. At each green dune and golden strand the ocean gives far more than it takes away. Long, violent breakers on these slowly shelving shores accumulate tons of calcareous shell-sand and heave it, waveful by waveful, ashore. This sand soothes acid peat like an antidote in venom. The sea is therefore 'loved' as well as 'feared' in the words of one of the islands' many poets, Iain Crichton Smith; it is not just 'monster' but also 'creator'.[1] In the dunes – whose unique flowering grassland born of shell-sand is known as machair – the sea stirs the most dazzling display of verdant growth imaginable. At the exposed edge, marram grass takes hold, its tight, spiky stems withstanding ocean storms and its seeds released only by gales. A few feet of marram can fix loose sand in place, and in its shelter the carnival begins. Red fescue, sand sedge, buttercup and primrose go through cycles of growth and rot which rejuvenate these sand-lands every season. Cloud clings to the high, peaty ground inland, while sun shines more often on these low, flower-clad expanses. Looking up from the kayak even at the end of summer, it's as if the foreground was painted in children's colours: a shock of vivid greens, yellows, pinks and orange. Behind is a sombre painting in night-time hues of brown, blue-grey and black (figure 4.1).

Wandering ashore through the machair onto dank rough moorland, the details are no less contrasting. September and October were the perfect months to see this. Where the rock holds sway, summer had gone. Slimy, nicotine-yellow smudges, once shining bog cotton, were peppered with crisp brown cadavers of tiny heather bells. Both mouldered in pallid clumps of sphagnum moss. These acid hills are no drier than the sea. Where the rock and peat reach the shoreline, vinegar meets brine. But sand dunes clung to summer: white mayweed, pink centaury and yellow trefoils still shone out, even when the grasses round them faded into winter. The disjunctions between the dark land and the bright are almost unbelievable in scale. Rich tropics and blasted Arctic wastes are separated by a broken drystone wall. And these contrasts have formed every stage of island histories.

Competition for the rich coastal land defined the push and thrust of skirmish, treaty and contract, whether made by locals with Edwardian landowners or between the sea civilisations of the Norse and Irish Celts. But not just conflict was driven by these patterns. Local ways of life responded too. Perhaps the most romanticised element of Western Isles history is the system of transhumance formed around the shieling: the stone or wooden bothy at the heart of summer pastures. Each summer, rich coastal crops, engorged by shell-sand, would be guarded from animal interference by leading herds to seasonal communities on the dark moorland interior. The bright nights on these wild moors, where the song and dance of ceilidhs filled still evening air, have long been evoked as symbols of premodern tradition, attuned to nature, in contrast to the antagonisms of modernity. The late-Victorian Celtic revival set a trend for shieling songs and poems: William Sharp, Marjory Kennedy-Fraser and Edward Thomas were among many who rehearsed such themes. Thomas linked the arts and crafts of moor-dwellers with the wild art 'painted by the wind'. In Kennedy-Fraser's Gaelic shielings, natural beauties – a dew-draped, sun-lit gentian and a 'white lily, floating in the peat hag's waters' – are like the eyes of 'Mhàiri, my beloved'. For William Sharp, as for countless others, long evenings on isolated moorlands primed tales of illicit trysts. The shieling was 'the hut of dalliance' and, in one Skye place name, Araidh na Suiridh ('the Bothy of Lovemaking'). The last evening of summer (*Oidhche na h-Iomraich*, or the night of the flitting) became for these storytellers the great set piece of song, dance and romance: the ceilidh to end all ceilidhs.

Twentieth-century ethnologists repeated these themes when they approached the Western Isles; Victorian assumptions concerning traditions entrained to nature died hard, as did voyeuristic images of happy primitives. A 1938 interpretation of shieling practices, penned by the ethnologist, E. Cecil Curwen, was even entitled 'The Hebrides: A Cultural Backwater'. Such articles belong to a time when shieling customs were giving way to standardised farming practices. As transhumance began the slow slide from living memory, folklorists and antiquaries took new interest in collecting recollections from the last generation to live the practice. They reinforced old themes of free immersion in pristine nature:

First night out on the moors was always strange . . . The smoored fire and her brood sleeping were the only familiar things . . . Round the bothy she could hear animal noises which gave her a sense of security. The silence was also occasionally disturbed by the cackling of grouse, the bleating of sheep, the splash of the jumping trout.[2]

When intensive farming practices proliferated, and landlords' efforts to 'rationalise' production banished communities from the land, many hillsides and coastal fields where low-intensity growing and grazing once took place were abandoned to the wild. Claiming peatland for farming requires the constant, back-breaking effort of kneading shell-sand through the peaty loam. Untended, such efforts unravel and bright lands fade to darkness: 'The hills that were green when I was a boy', one local told geographers in 1961, 'are now black.'[3] Rushes and flags conquer the wet land while heather reclaims solid clods. As I travelled south, I slept on islets unlike any I'd stopped on before. Some were swamped with rampant foliage because, despite their staggering fertility, they've been ungrazed for a century. Small birds, voles and rats delve narrow tunnels through the densely woven grasses, and animal tracks are neither sheep nor rabbit but otter.

I spent my nights listening to roving creatures as I gazed up to dazzling starlit skies and brushed rogue springtails or spiders from my sleeping bag. Contrasts of land and sky made sunsets and dawns unnervingly beautiful, and dusk rainbows abounded. It's the most terrible paradox of these islands that, despite the power of this dramatic landscape to shape history, it's often the richest and most delightful spots that are most conclusively abandoned. Clearance in this region involved the devotion of machair to sheep, and the founding of new townships in barren districts of the acid land. Nowhere else in Britain have the processes of history more fully sundered the usage of land from that which seems most natural. The best-known results of this disjunction are the coffin roads. When nineteenth-century landlords cleared ancient communities from the Atlantic coasts, all that remained to link the present with the past were cemeteries scattered in the machair. The sites to which people were cleared lacked sufficient soil even to cover a corpse, so communities escorted their dead over miles of rough moor to be buried at ancestral sites. I saw these cemeteries every day, some on tiny islands like Little Bernera (Beàrnaraigh Beag) where no structures but the chapel and its graves survive (figure 4.2).

Where romanticised visions of the shieling coloured many accounts of Western Isles life from the late Victorians onwards, it's the darkest spells in history that tend to dominate accounts by visitors today. The coffin roads and coffin ships, in which thousands of unhappy islanders were forced to the Americas and Antipodes, each have their gripping histories, as do the actions of aristocratic island-owners who practised the arts of imperial exploitation in the Hebrides as well as on the Chinese victims of their trade in opium.

Yet the most striking stories to emerge from my coastal journey were entirely different from either the sentimental memories of shielings or the grim tales of exploitation and abandonment. They involved phenomena I'd not anticipated until I was afloat. On my first day at sea I encountered one of the most astonishing resources I've ever made use of: an incongruous coastal archive that brimmed, un-nostalgically, with optimism, courage and excitement. This was a collection that drew in more than half the globe yet remained unmistakeably, idiosyncratically local.

On the morning I reached the archive I had, quite deliberately, told no one in the Western Isles' northernmost district, Ness, of my need to edge around their infamous headland. This is the rocky, windy Butt of Lewis where the swell of the Atlantic meets the tides of the Minch. Later, once I'd landed, I was regaled with tales of shipwreck and sea death that might well have sapped the conviction I required to grit my teeth through danger. But I'd learned by now that local knowledge is terrifying, and anyone who heeded it would never kayak anywhere. Children raised on North Atlantic shores are taught to view their own seas as uniquely treacherous. Every fast tide becomes the fastest tide. Every seam where swells converge becomes the most dangerous spot in Scotland. Yet there's something particular about Outer Hebridean respect for their unquiet ocean: when I'd arrived in Lewis's east-coast capital, Stornoway, people were far more sceptical of my journey than Shetlanders or Orcadians had been. The response I now met was less 'what a wonderful thing to do' and more 'rather you than me'.

I soon saw why. An old Lewis saying – 'the sea moves faster than the wind' – rang true as I rounded the Butt. Skies spoke the language of summer: clean, blue and close to calm. Yet the sea seemed to contain the violent energies of winter in its gliding, grinding waves. Coastal kayaking is different here from in the Northern Isles because land drops away slowly rather than suddenly, leaving shallow inshore waters and multitudes of reefs. These obstacles for the incoming ocean transform swell into tall breakers that can run huge distances before erupting on the shore. Within ten minutes of launching I watched spume climb cliffs, reaching so high that the water from each wave barely began its descent to the sea when the next bombardment hit (figure 4.3): the result was a great salt Niagara that foamed just half a mile from a massive tidal stream known locally as 'the river'.

I knew then that every landing between the networks of boat-shredding rocks which guard each beach would be a lottery. Tides, till now my biggest threat, were of little concern in comparison to the rolling swell that was channelled into bays as breaking sea. For centuries, the rigging of fishing boats on this jetty-scarce coast had been adapted for launches through surf: the thirty-foot boat called the *Sgoth Niseach* has a long, high yardarm that holds the sail's low edge above the breakers. But even with such precautions, it's impossible to launch from western shores through much of the year. Most creeling and fishing is thus an east-coast affair: few people work the autumn Atlantic waters. Gulls, auks and fulmars were as absent as fishermen now the summer had ended: nothing moved at the Butt but the sea, and me, dizzyingly, on it.

In every fracture of this scarred coastline there are stories, preserved precariously in local memory. I embarked from a cove below the scant remains of a chapel named for the most venerable local saint. St Ronan, whose name means 'little seal', was a seaman to whom God gave as a ship a great leviathan of the ocean, the whale-eating *cionaran-cro*. A mile west, a tall, inaccessible wedge torn from the cliff face, Luchubran, is the site of another chapel ruin and of strange tales of ancient Spanish pygmies who dwelt on Lewis for five centuries, before, in the year 1, they were harried to this steep rock by 'large yellow men' from Argyll. A mile further and the cliffs give way to sand and surf: this is the beach from which the 'little seal' embarked astride his sea beast to answer the deity's call on a remote isle, Ronaidh an t'haf ('Rona of the Ocean'; a more telling name than the English 'North Rona'). Rona is one of innumerable ocean fragments that mirror at sea the ragged indentations of the Lewis coast. From the Monachs, Flannans and St Kilda in the west, Sula Sgeir and Sula Stac in the north, to the Shiants in the east, the Outer Hebrides are the gravitational centre of a grand, imposing ocean world where medieval saints built cells among seabirds, and seabirds built nests in the skulls of saints.

I'd travelled only five miles when I reached St Rona's beach, but it was clear that without a *cionaran-cro* I couldn't battle this ugly sea for long, so I braved the surf and was delivered, tumbling to the shore. Dragging my boat into the machair beneath a wide cemetery, I wandered into one of the richest historical regions in Britain: the fifteen townships of Ness.

For a millennium between the sixth and sixteenth centuries, Ness was the pivot of flourishing ocean cultures. It was the gateway to the sainted skerries: ancient chapels line Ness shores, still haunted by the sounds of the spiritual sea. The most famous, at Eòrapaidh, stands intact above the scattered crofts of the township called Fivepenny. Yet there's no written evidence of Ness history from the era of the oldest chapels: the many stories that survive are products of the region's unrivalled oral custom. Such stories persist because they explain how Ness became the way it is, and historians of the Gaelic-speaking world (the Gàidhealtachd), such as Domhnall Uilleam Stiùbhart, have begun to show how they offer opportunities to develop new methods for writing British history (a field long stubbornly reliant on texts).[4] Ness was also a hub from which trade travelled east and south: from the fortified port of Dùn Èistean the great Macleod clan oversaw the movement of goods and people across the many seaboards they controlled.

But every century from the sixteenth onwards struck a blow to island independence. First, James IV undid the power of the sea clans; there followed the unruly *linn nan creach* (the time of raids). In the seventeenth century, harried by agents of James VI, Dùn Èistean fell from use, and Lewis ceased to be a seat of power. A century later, the triumph of Hanoverian kings over the Scottish Stuart line moved power still further south and increased the determination of Lowlanders to break the clans conclusively. By the start of the nineteenth century the Macleod lords were gone, replaced by mainland agents of the British Empire (Mackenzies and Mathesons) who subjected Lewis to the profiteering with which they ravaged the globe. Communities were economic assets to be moved between sites like cash between investments. The gravitational centre of island life moved east, away from the Atlantic: traditional seats of coastal power – such as Ness and Ùig – gave way to the growth of Stornoway. When incomes from kelp collapsed, the priorities of landlords turned conclusively from the shoreline: for the first time in Lewis's history, the strand was seen as valueless.

Yet even amid immense poverty and exploitation, Ness somehow found ways to flourish. The Gaelic language and local life survived the centralising onslaught. Indeed, the peak of the Industrial Revolution, between 1755 and 1840, had seen such growth in the Outer Hebrides that population expansion (98 per cent on Harris, 141 per cent on North Uist and 211 per cent on South Uist) carried the number of Gaelic-speakers in Scotland to its highest figure in history.[5]

1840 was a turning point: after decades of roughly parallel evolution, the fates of Lowland Scotland and the Gàidhealtachd diverged. Where Glasgow entered its most productive era, attempts to exploit the natural resources of the islands failed. The efforts of landlords to 'improve' agriculture generated famine and disease but little profit, before landowners sought to ease 'congestion' and reduce their responsibilities by demanding emigration. At the beginning of the twentieth century, Ness had two faces: life expectancy was just forty-nine and destitution widespread, yet this remained a place of rich stories and many industries. Small factories made lemonade and biscuits; workshops made textiles and shoes (at least seven cobblers are recorded in Ness around 1900); Ness ships continued to carry such goods across centuries-old sea routes, and Ness men and women remained lynchpins of massive industries such as herring.

Ness had survived a series of seemingly insurmountable crises, always able to find new outlets for economic and cultural invention. Yet the early twentieth century saw crushing blows and the onset of a true dark age. The First World War was the biggest tragedy in Ness history: a moment that still dominates local memory.[6] With its strong seafaring customs and a long tradition of sending hale young men to police the streets of Glasgow, Ness was a prime recruiting ground for the naval reserve. Nearly 1,000 Ness men and women served in war; a quarter were lost in action. This was a similar scale of loss to that of many other communities. Yet the tragedy was magnified by the infamous *Iolaire* disaster when, on New Year's morning 1919, an Admiralty yacht carrying 300 island servicemen struck rocks just metres from port and, in sight of crowds waiting to welcome them home, 200 drowned. Ness entered a depression – economic, demographic and emotional – from which recovery would be achingly protracted.

In 1918, Lewis was bought by Lord Leverhulme who labelled local crofting a 'medieval' form of land mismanagement; pre-war plans for aid were abandoned. But Leverhulme saw the seas as a site for industrial methods. People of Ness returning from war found their waters and ports unrecognisable, plied by English and east-coast trawlers who channelled catches from the Minch directly south. In Ness, they complained in 1923, there was now 'no fish obtainable . . . there were not twenty lings landed at Port-of-Ness this season, where hundreds of thousands were cured annually once upon a time'.

The herring industry soon unravelled too, and any chance of Ness being saved by the sea died with it.

By mid-century, the Outer Hebridean islands were the most depressed regions of Britain. They had the highest unemployment and the least chance of retaining young people. This was when studies of the area described a 'cultural backwater' and even many islanders began to believe the narratives of failure. 'Traditional' was now a term to contrast unfavourably with 'modern'. In the face of economic crisis, a bright future was one in which mainland, urban culture ameliorated the isles. Everything distinctive about Lewis was seen as an embarrassment that impeded regeneration and left the island floundering in the past. One Ness resident told me that once a cash economy was in place, crofting seemed unsustainable. He described 1940s women in their early teens wandering to the port seeking work in mainland hotels; he described his own act of setting off for Customs House, aged seventeen, to join a huge merchant vessel in which hundreds of youths set out for South Georgia. With movement on this scale, the transgenerational transfer of stories, on which Ness identity and culture is constructed, had little hope of surviving the century.

The crusade against Gaelic embodied all that was worst about the era of industrial modernity. Clearances divided Gaeldom, so that a third of Gaels spoke their tongue in places with names like Skye Glen and Glencoe on the Atlantic coasts of Canada or South Africa. The proportion of Scotland speaking Gaelic slid from a majority to as low, in 1951, as 2 per cent. and these few increasingly dwelt in tiny pockets of the former Gàidhealtachd: all or almost all were poverty-stricken coasts of the Atlantic edge.

Central to the language's impediments were the Education Acts of the early 1870s. For much of the British Isles these celebrated Acts (covering England and Wales in 1870 and Scotland in 1872) heralded a bold new dawn of educational opportunity, but for anywhere with different educational needs from urban England, the new standards were disastrous. They were, to the educationalist Farquhar Macintosh, 'the most serious blow that Gaelic suffered, more than . . . the clearances or even . . . the terrible toll that was taken of Gaelic-speakers in the First World War'.[7] The Scottish Act assumed that speaking Gaelic interfered with the proper learning of English and so held back the development and opportunities of any child unlucky enough to be a Gael. Overnight, schools born of Gaelic communities became alien, English-speaking units. As Christine Smith observed when visiting rural Lewis in the 1940s, the most mundane features of mainland textbooks – railway stations, lamp posts or cricket bats – were as unrecognisable and irrelevant to children as guga would have been in Clapham. Iain Crichton Smith wrote that in a short bus ride from school to home 'I moved between two worlds'; he had never left Lewis but had never read books that didn't come from England or America.[8] The result was that, where English children found the culture of the home buttressed and confirmed in school, the first intellectual training for Gaelic-speakers served to 'undermine, to weaken and to harass' the home's heritage.[9]

Statistics for the Western Isle of Eriskay in 1890 drive home the split of school from home. Of 5,821 Eriskay inhabitants 5,532 spoke Gaelic and 3,430 spoke only Gaelic. Just 289 spoke no Gaelic, but three of that tiny Anglo band were the schoolmaster and his family, shipped in by the school board from Birmingham.[10] The Education Act was followed by a bombardment of anti-Gaelic propaganda. Simon Laurie, professor of education at the University of Edinburgh, insisted that a child taught Gaelic had been 'miseducated – in fact, cut off from being a member of the British Empire altogether'. In 1878, the leading Edinburgh publisher and politician, William Chambers, published two caustic articles on 'The Gaelic Nuisance'. Because Gaelic-speakers didn't have English books and newspapers, he insisted, they must 'vegetate between vague legends and superstition'.[11] To learn Gaelic, he stated, was to remain ignorant: to wipe out this abomination required 'moral courage in the face of popular prejudice' but he saw this crusade gaining momentum around him. In the same year, the *Scotsman* pronounced that the Lewis people, who continued to try and live through Gaelic, were 'totally unprepared for the good that the Act is expected to do to them'. In the face of this onslaught, strong Gaelic-language activism gained some concessions and inspired a revival of the language in print, yet the breach between the Gaelic home and the artificial English schoolroom persisted till the 1980s.[12]

Language was the most profound victim of nationalised education, but Hebridean histories were also casualties. An education lauding mainland monarchs and remote English cities left children ill-equipped to comprehend and value local life. As Crichton Smith put it, 'the island seemed to have no history [because] it never occurred to anyone to tell us'. Lewis felt, as a result, like 'a hard bleak island which did not reverberate when one touched it with one's mind'. Children were educated as though Lewis was just a stepping stone to more important places: 'slowly', wrote one activist in the 1970s, 'the folk of Ness see the viability of their Gaelic community disappearing . . . we have been educated for one purpose – to be shipped across the Minch to find work'.[13]

The building I arrived at when I wandered into Ness had witnessed all these changes. As the school for several Ness townships it once had 265 pupils but, after decades of declining numbers, had closed. In 2011 it was reopened as an archive, museum and café. I'd heard this centre was run by the local history society (Comunn Eachdraidh Nis) and that this group was now, remarkably, the biggest employer in the region, so I was intrigued to see their work.

Salty and damp, I stepped into an archive which was so bustling with people that it felt entirely unlike the quiet emptiness historians anticipate of urban equivalents. Locals switched effortlessly between

Gaelic and English as they conferred with each other and with the Canadians and Australians brought to the resource in search of Ness ancestors. Over and over again I heard visitors and locals find points of commonality in relations who died a century ago, or in Ness trades now gone. Within an hour the centre's guiding star, Annie MacSween, had ushered me to another room to speak to local retirees about my kayak journey. One of them told me how the villages of Ness had migrated inland when the road was built. Buildings that once faced the ocean had been rebuilt 200 yards uphill to face the cart track instead: this community's perspective on the world had been transformed. An hour later, MacSween was helping me find documents to learn of the historical society's early work. But only later, when I began cross-referencing texts from Ness with others in Stornoway archives, did I discover the scale of MacSween's own contribution to Lewis life. This wasn't just a matter of dedication to community, but of a visionary approach to both history and social regeneration.

In 1970, the first stirrings were taking place of processes that offered hope to coastal regions such as Ness. One important step was the reorganisation of local authorities: rather than being split between faraway mainland councils in Dingwall and Inverness, the Western Isles became a unified body and public servants could operate from Stornoway rather than living far across the Minch. This reverse of the island 'brain drain' assisted the attraction of resources from private as well as public bodies. The Bernard van Leer Foundation, established in 1949 by the Dutch oil magnate Bernard van Leer (1883–1958) to aid every area in which his outfit operated (from Colombia to Finland via Ness), began to see the need to avoid central, patrician provision in favour of control by people it labelled 'free-wheeling activists' with 'a finger on the pulse, who can respond quickly to needs as they emerge'.[14]

But even this enlightened organisation couldn't have foreseen how MacSween (then called Annie Macdonald) would use their funds.

She pioneered one of the most successful job-creation schemes in the history of the Scottish coastline, not by founding a fishing co-operative, a credit union or a transport company but by recruiting five unemployed young people and a project leader, Agnes Gillies (tempted back to her native Ness from Aberdeen), to collect oral histories from locals. At first this novelty looked eccentric. In a 1979 interview, Macdonald noted the concerns of her critics: 'perhaps they thought it wasn't the best way to spend public money. Maybe they thought the past was dead.'[15] They had reason to be sceptical: Scottish studies in this era took folklore from the Gàidhealtachd but showed no ability, nor even ambition, to feed back into the culture it collected from.[16] Yet the legitimacy brought by official funding made it possible for the Ness group to begin a historical scheme crafted to stimulate 'the people of the Western Isles to perceive their own community more clearly'.[17] The possibility of a constructive cycle, where society made history and history made society, was born.

Nothing could be achieved economically, Macdonald had realised, until the narrative in which Ness people placed their lives was turned around. She and her two co-founders soon gained access to the major historical institutions of Edinburgh. As they wrote in their report of a 1978 tour of these centres, official funding gave

> an element of credibility to our visit in that it gave the impression, not of a group of eccentrics dwelling in the past, but of a versatile young group looking to the past for positive elements carried into the future. Indeed, more and more, the survey in the course of establishing an archive of local history is also proving to be a means of establishing our identity as a Parish.

And the group's purpose was simple: by recovering the herring girl, small crofter and Gaelic singer from 'the enormous condescension of posterity' they aimed to show that mainland pasts taught in schools were not the only worthy histories.[18] The historical society was like shell-sand worked through acid bog: an antidote to the idea that Gaelic culture was intrinsically inferior.

Macdonald's six workers toured the district. The '*muinntir an eachdraidh*' ('the history folk'), as locals labelled them, were soon familiar presences beside peat fires throughout the townships. They collected photographs and organised them into themed collections. They recorded Ness singers performing local songs and sold 400 copies of their first cassette. They presented these projects as opening up resources for community discussion. But the early history of the society also included unique endeavours which mixed historical research with immediate practical purpose. These were schemes that no office-bound autocrat on the mainland could ever have imagined. *Leabhar nan Comharraidhean* was a unique directory of local sheep earmarks, built on genealogical research, and profoundly helpful in the control of free-roaming Ness livestock. The society's *Leabhar Na Fon* included Gaelic patronymics of every telephone holder in Ness, facilitating communication in a place where a small range of names applied to a wide range of people.

The coastline played a profound role in these activities: early documents indicate that restoring emotional investment in what had once been Ness's most productive resource – the sea and shore – was high on the activists' agenda. The first talk given at a society meeting concerned the unique Ness privilege of collecting gannet chicks from the ocean rock of Sula Sgeir: the local newspaper noted that 'this proves a very interesting lesson for all the Niseachs present who have been eating the guga for many years, but had no knowledge of how or where it was caught'. The first exhibition featured early maps, exposing the memory of residents to the rich Gaelic nomenclature of coves and headlands in which old uses of the coastline were embedded. The first publication of photographs was *The Herring Girls* (1988), in a lavish edition printed in Gaelic on the island. And the first major fundraising initiative led to the salvage and renovation of a traditional boat, the *Sgoth Niseach*, which was a veteran of the lost line-fishing fleet. The records of the fundraising show joyous disbelief as donations poured in from around Scotland and from Ness émigrés on four continents.

The medium, however, was as important as the message, and reports of early events are emotive things. The twelve-strong committee was young – some in their twenties, at times with an average age in the mid-thirties – and such excitement surrounded the occasions when local people gathered to discuss their heritage that reports of committee meetings feature phrases such as 'before the meeting ended, it was more like the "tigh cèilidh" of old'. The early photographic displays were emphatically for locals rather than tourists, with only Gaelic texts; but they were also able to generate publicity – the first exhibition, for instance, was covered by the press including the BBC. The impact was immediate and profound. As one article put it, the exhibit 'brought people together in a new way. It generated . . . the sort of enthusiasm for action which had previously been lacking.'[19]

Ness's fit of historical fever proved infectious. By 1992 there were fifteen 'Comuinn Eachdraidh' in the Western Isles, their voice in public affairs aggregated through Caiderachas Eachdraidh an Eilein Fhada (the Lewis and Harris Federation of Historical Societies) which could assist in such projects as providing island-wide resources for schools. Today there are closer to thirty history societies, knitted into a network under the auspices of Tasglann nan Eilean Siar. Annie Macdonald and her peers in the 1970s took part in nothing less than a revolution in island life. It was only much later that celebrated reports, such as Michael Corbett's study of education in a Nova Scotia fishing town *Learning to Leave: The Irony of Schooling in a Coastal Community* (2007), would codify the issues with which these societies had been dealing for three decades.

Many of the history societies are sited on Atlantic shores. After leaving Ness, I found myself dripping at their doors again and again. Although all these centres fuse history and language with attentiveness to the relations of life with local landscape, their diversity is remarkable. Some, such as that in Taobh Tuath (Lewis), focus almost exclusively on genealogy; others document in detail the history of a local trade, such as Harris tweed; yet others commemorate individuals such as the extraordinary South Uist artist of artefacts found on coast and moor, Angus MacPhee, known as the silent weaver. By the time I reached Barra, many societies were reducing their activities for the autumn, so that their busy custodians could take their own holidays, but without exception someone was willing to sacrifice their time for me to sift their holdings. In historians such as Mairi Ceit MacKinnon, custodian on Barra, I met other vital figures in the building of bonds between historical research and community cohesion. MacKinnon let me sneak into one side of the archive late on a Monday night while, on the other, she held a meeting to discuss, in soft Barra Gaelic, the island's contribution to the upcoming festival of Gaelic culture, the Stornoway Mod.

Today, the infrastructure of Gaelic community exists on a scale unimaginable half a century ago. Where industrial developments from 1840 to 1970 largely worked against island economies, those of more recent years have done the reverse. Where roads and rail served to isolate the islands from new networks of trade and power, the communications revolution has reintegrated them. Island broadcasting and publishing flourish: presses such as Acair (based in Stornoway) ensure that island literature reaches far beyond the Hebrides. Arts centres such as An Lanntair run bilingual programmes to guarantee that, where islanders once fled to the mainland, it's now mainlanders who flood to Lewis for culture and entertainment. However, perhaps the most powerful innovations of all are those that undid the educational travesties of the twentieth century. The

University of the Highlands and Islands was founded in the 1990s, with the historian Jim Hunter as one of its leading figures, to make certain that Gaels aren't forced to leave the Gàidhealtachd to receive a first-rate education. One of its thirteen colleges now sits in the old Lews Castle in Stornoway from which mainland lords once aimed to anglicise the islands.

But opinions differ sharply on the current condition of Gaelic. Although younger islanders I spoke with talked of a thriving language, when I visited veterans of the 1970s campaigns they weren't so sure. Everyone expressed the importance of Gaelic education: bilingual children, one noted in a wonderful phrase, 'have two wisdoms on the world'. The expansion of media and of Gaelic-teaching in Glasgow and Edinburgh is, that islander continued, perhaps leading to some 'groundswell of feeling that Gaelic is going somewhere'. But the point he wanted to drive home was that there's still no place where it's easy to raise a child in Gaelic: it's 'a political act . . . almost every move [parents] make has to be against the tide . . . as a living, family, normal, unconscious language it is . . . very weak'.

As I moved from low Lewis on to the mountainous coast of Harris, the wind turned easterly, so flattened the swell and eased my passage. I spent a night between lazy-beds in an extensively worked glen now many miles from any village or working croft (figure 4.1). At dusk a golden eagle preened feet away and I collected a huge downy feather from where it perched. At dawn a sea eagle flew close by while I was peering down at jewel-like sea gooseberries washed up along the sands. Lazy-beds, *feannagan* in Gaelic (a term that carries none of the bizarre condescension of the English name), are themselves a modern, introduced technology. Potatoes were the miracle food of the early-modern coast. Visitors to the Scottish and Irish islands in the eighteenth century describe a population living in abject poverty but who were tall and strong. These tubers carried far more nutrients and were more efficient, damp-resistant and voluminous than the grain on which islanders had previously relied. For two centuries after the introduction of the potato, the coastline of the islands was a scene a Peruvian would have recognised. Thin soil was piled into raised beds, allowing a double thickness for planting, and permitting drainage along the intervening furrows, so that coastal land was shaped like an artist's imitation of the rolling sea. At the beginning of the nine-teenth century ambitious landlords, following the ideals of new, Enlightenment, agricultural sciences, sought to consolidate lazy-bed patches into large fields to be ploughed rather than dug. This inten-sive system was what allowed the spread of blight in the 1840s, generating famine and disease throughout the Gaelic zones of Scotland and Ireland and precipitating the demographic catastrophes that have shaped these coastlines as we see them.

These features of Western Isles coasts were my first indication of the problems modern farming introduced. Soon, I crossed the Sound of Harris into Uist. I slept on a tiny island, Vallay, beneath the ruins of a large house built by the archaeologist whose work brought these islands to the attention of historians worldwide, Erskine Beveridge (figure 4.4).[20] Vallay sits amid wide sands and shallow seas where the sounds can be strolled at low tide. Many walkers, including Beveridge's son, George, have died mistiming crossings. These Uist sands are so voluminous and mobile that in a storm of 1756 the houses of the township of Baleshare were buried to their roofs in beach. But the payoff for this sandy treachery is the finest and most expansive machair in the world.

Today's best machair has been shaped by centuries of low-intensity human use, but 'improved' farming brought many threats. Historic, small-scale crofting led to a patchwork of habitats where small strips of land were grazed, grown and left fallow every year. Mammals, birds and insects flourished among the crops and wildflowers: corncrakes and short-eared owls bred here as in no other farmland. As tenants were evicted and farms consolidated, the jumbled land was replaced with large clean units subjected to a single use. In the hope of enriching grazing, grass was sown that choked the flowers and lacked the long roots that could hold the dunes together. Hay was gathered early in the year, scything wildflowers before they'd dropped their seeds and stripping habitats whose chicks hadn't yet fledged. Chemicals superseded seaweed as fertilising agents, transforming the mineral content of this fragile land. Without a thriving machair, no expense or technology would keep this landscape fertile.

It took decades from the realisation the dunescape was dying for solutions to be found. In ways that couldn't have been predicted in the 1970s, the answers related to the resources of the Comuinn Eachdraidh. Twentieth-century farming, which made a roofless factory of the fields, was unique in its relentless focus on the present. Like the people of Ness, those who worked the land in Uist found that looking to the future meant recovering the past. For centuries, this landscape had been productive without the threatening novelties, yet the skills and routines of the machair crofter had been slowly falling from memory. Research into machair management involved talks with ageing crofters and the collection of oral histories. In 2010, a project called Machair Life put this research to use, encouraging traditional crofting practice as the only form of sustainable production in this landscape.

The case of machair crofting is mirrored across the Western Isles: in the last decade, small entrepreneurial ventures have made many new old uses of the land. When a distillery was founded as a social venture on Harris, it recruited a local herbalist, Amanda Saurin, so that traditional resources – meadowsweet, wild thyme and kelp –

might be the botanical basis of its trend-setting gin. The rich culture of tweed making – its colours recruited from flora and its songs preserving island history – was once endangered, but the output of island handlooms and the affection felt for the tradition has grown so dramatically in a decade that tweed once again appears on the catwalk and in high-fashion furnishing. The restoration of pride in the local, and the destigmatising of tradition, has inspired a many-layered revival.

One aspect of this Hebridean renaissance has received more attention than others. Many readers of nature writing will be familiar with the concept of a 'Counter-Desecration Phrasebook'.[21] This originates on the west coast of Lewis. In 2010 a multinational energy consortium, AMEC, unveiled plans to turn a large tract of island into the largest onshore wind farm in Europe. The project's supporters described the region as 'a vast dead place' devoid of nature and culture. One Lewis resident, the psychologist and historian Finlay Macleod, orchestrated the resistance to the wind farms in ways that echoed the historical values of the Comuinn Eachdraidh. The long threads of Gaelic tradition in this landscape became the means of proving that culture and nature thrived in the moorland. 'What is required', Macleod insisted,

> is a new nomenclature of landscape and how we relate to it, so that conservation becomes a natural form of human awareness, and so that it ceases to be underwritten and underappreciated and thus readily vulnerable to desecration. What is needed is a Counter-Desecration Phrasebook.[22]

Taking up this call to arms, one Lewis activist, Anne Campbell, collated a rich historical vocabulary of Gaelic environmental language, *Rathad an Isein* (*The Bird's Road: A Lewis Moorland Glossary*, 2013) endeavouring to show that Gaelic was a culture of nature whose ways of ordering the world could resonate with radical environmentalism today. In current environmental politics, when every wind turbine is celebrated as a great acceleration of the flight from fossil fuels, this episode shows that it's worth taking note which communities tend to pay the price, in damage to their ecologies and cultural heritage, for the colossal energy consumption of cities.

At the same moment, Britain's leading nature writer, Robert Macfarlane, was seeking new ways to bring cultures and histories into his prose. A stinging critique of his writing, penned by the Scottish poet Kathleen Jamie, had accused him of 'quelling our harsh and lovely and sometimes difficult land with civilised lyrical words'.[23] Macfarlane set about answering this critique by building rich new histories into his work, and in Macleod's idea of a Counter-Desecration Phrasebook, he discovered the concept with which to transcend the limits of his previous writing. By the time of *Landmarks* (2015) the lyrical but unpolitical prose of early books such as *The Wild Places* (2007) was gone, replaced by a far more purposeful fusion of living histories with active nature. Macfarlane's intervention gave Macleod's idea a vast global audience, but the Western Isles gave to Macfarlane, and to modern nature writing, something equally important. What Macfarlane picked up was a product of the active practice of history that had brewed in Lewis since 1970. The Ness cultural revolution now gently shaped the attitudes of millions who'd never heard of Ness.

Many historians like to think they offer hope in the present. The question 'what is the nature of the present?', these historians argue, can be answered only comparatively, and the past is our only real source of comparison. It shows us there are other ways to live than those practised today. The past is full not of dead things but of unfinished business: germs of fruitful routes as yet untravelled. Every coastal ruin whose living cultures were once steamrollered by the homogenising logics of industrial capitalism is a site at which the possibilities for an escape from those logics can be entertained. Knowing the past events – more often odd accidents than the inevitable workings-out of well-made plans – through which our societies became what they are is a crucial prerequisite to escaping visions of the present as monolithic and invulnerable to change. Such historians are often concerned with the analysis of social inequality in the past and present, and they frequently quote a phrase from the Italian thinker Antonio Gramsci who demanded 'optimism of the will, pessimism of the intellect' from those who wish to change the world. He meant that clear-eyed analysis of social problems should go hand in hand with unshakable belief that all such problems can be overcome.

Ecologists often think in similar ways to these historians. Their knowledge of the richness of our ecosystem in recent history creates an intense awareness of present degradation. To be an ecologist, wrote the American environmentalist Aldo Leopold, is to 'live in a world of wounds'. It is also, however, to have at least some knowledge of first aid and a commitment to utopian thinking: ecological knowledge provides the ability to see the mechanisms, however far out of reach, by which current crises might be reversed.

However, for both historians and ecologists 'pessimism of the intellect' is more easily sustained than 'optimism of the will'. Few find themselves genuinely empowered to turn analyses of past or present into better futures. In this sense, Comunn Eachdraidh Nis, the Counter-Desecration Phrasebook, and the Uist crofters are inspirational: they have done, on minimal resources, what much larger historical bodies can often only dream of. This is why the Western Isles should be a site of pilgrimage for anyone who wonders what history can do or how historical thinking can be made to matter.

My last days in the Western Isles involved some of the most dramatic kayaking in the north-east Atlantic. Surprisingly – miraculously – autumn refused to arrive. September gales were still awaited at the end of October; rainfall was lower and sunshine higher than in any autumn for decades. I passed south of Barra during a blazing sunset when the swell was ten feet high and rippling gold in the dying light. Flocks of eider, hundreds strong, lurked unseen between the burnished seas so that long descents down waves left me running through their midst. In a mighty whirr of wings they'd rise in unison and make for the distant isle of Mingulay (figure 4.6). I too was hoping to be among the glorious islands south of Barra – 'the nearer St Kildas' – by dark. Sandray (Sanndraigh), Pabbay, Mingulay (Miughalaigh) and Barra Head (Beàrnaraigh) are among the wildest spots in Britain: huge rocky mounds that slope from low eastern shores to intricate towers of Atlantic-pounded cliff. These islands are both an archaeologist's dream and ornithological heaven: count-less rare species, in the course of migration, alight in Neolithic field systems that survive like nowhere else in Britain.

But the Barra Isles are also a reminder that not everything lost in the coastlines' darkest days can be recovered, however dramatic the island renaissance. The sky was black by the time I clattered into a corner of Pabbay and pulled myself ashore. I was among sparse remnants of a complex of ruined crofts, huddled round the island's stream. In the eighteenth century, three families lived here, numbering sixteen people. By 1851 the three households numbered twenty-five. There was never a school on Pabbay, although records show a 'tutor' living with one of the families: a fourteen-year-old boy from Barra. By 1861, for some unrecorded reason, the population had been replaced by incomers from South Uist who built a large gabled house apart from the clustered crofts. These were people seen (by themselves and others) as foreigners with little experience of a life dependent on the sea.

The Uist incomers are remembered as prodigious distillers of illicit whisky. In the morning I climbed down to Sloc Glasnich – a steep Atlantic inlet where the illegal still was hidden. Islanders also exported tallow and kept much livestock: forty-six cattle and thirty-seven sheep were recorded in 1883, while a pony was left on the small neighbouring isle of Lingay (Lingeigh) to be used when peat was cut there (stills, many moonshine-isles attest, gobble peat prodigiously). Pabbay's cliffs house few seabirds for food, but islanders took guillemots from sea stacks and harvested vast quantities of feathers: consignments of up to fifty kilos were shipped for sale to Greenock. In 1888, the islanders had a herring ship and three small lobster boats, but these are not the kind of seas where vessels have long lives, especially in the hands of those unaccustomed to oceanic living. First, the herring boat silently disappeared from the records, then the number of small boats fell, until, on May Day 1897, disaster struck: the island's only boat was sunk, drowning five. Soon after, Pabbay was abandoned forever. I found just two other structures on my circuit of the island. One was a large concrete square surrounded by flat stones: a salt box around which fish would've been laid for curing. The other was a mystery at first: only later, when I met the artist Julie Brook in Stornoway did I discover it was the remains of one of her fire-and-sea artworks and that she was perhaps the only person for a century to have spent long spells on Pabbay.

Besides the island warden, based on Barra, Pabbay has visitors only rarely. The few who arrive are often enticed here by whispered rumours in distant bars that these cliffs host the best climbing in Britain. Sweeping structures in the sea crags, such as the Great Arch, are riddled with stunning routes that challenge the most intrepid, with names like Prophecy of Drowning (climbing grade E2 5c) and Child of the Sea (E5 6b).[24] But one route on the Great Arch isn't named for the ocean; its title, the Priest (E1 5b), denotes the aspect of Pabbay history that brought me to the island and tied off another thread through my Western Isles experience.

Since starting my journey, I'd landed not just on one Pabbay, but a host of them. I'd stopped at Pabay, Pabanish, then alternating Pabbays and Paibles until this final stop beneath Barra. By turns these names mean 'island of priests' (Pabbay/Pabay), 'settlement of priests' (Paible) and 'headland of priests' (Pabanish). But even the Paible settlement sites are on small islets of staggering beauty, from Little Bernera (Lewis) to Taransay (Harris). I'd seen many 'papa' sites in Orkney and Shetland, such as Papa Westray, which carry the same meanings. Such names remain an obscure phenomenon. They belong to a Nordic nomenclature, so seem to reflect the encounters and experience of Norse seafarers. The only early textual description of the *papae* of these islands is a twelfth-century history of Norway that clarifies nothing:

> the *papae* have been named for their white robes, which they wore like priests. But, as is observed from their habit and the writings of their books abandoned [in Orkney] they were Africans adhering to Judaism.[25]

The source goes on to say that the Norse despatched great fleets and destroyed the *papae* wholly.

This account plays a minimal role in today's interpretations. Pabbay names are treated as evidence of encounter: key sites in the expansion of Irish Christianity and its eventual coexistence with the southward-spreading Norse. In the hands of saints such as Ronan, Columba and Brendan, the Celtic church had expanded north from Ulster, moving by Atlantic routes (with two small exceptions, the Inner

Hebrides lack the Pabbay place names). Such sites mark, therefore, the identity of the Outer Hebrides as an archipelagic frontier between societies.

In the twentieth century, the *papae* were seen as holy men seeking a 'desert place in the ocean' and choosing these island sites because of their remoteness. But only a modern landlocked culture could misinterpret coastlines quite so drastically. Two things demolish that interpretation. One is the centrality of these islands to premodern trade routes. The ocean lanes now sometimes labelled 'sea roads of the saints' were general lines of travel for goods, ideas and people: these seas were society's arteries. In later centuries Macleod lords set up home near two old Pabbays; historians long wrote of these lordly manors as sitting at 'extreme outer limits of their territories'. It was an interpretive leap made in the 1980s that stopped seeing 'second homes' (where lords 'got away from it all') and recognised these islands as crucial sites for ruling sea realms.[26]

The second reason to reinterpret the *papae* sites is that these machair-rich islands are anything but 'desert': they're often the richest landscapes in their region. In mid-September I'd stopped at one of the most beautiful spots I've ever been: the isle of Pabbay Mor (Pabaigh Mòr) at the head of West Loch Ròg. As the last familiar mark for Lewis folk embarking on Atlantic journeys, this is an island rich in song and verse.[27] West of Pabbay Mor's abandoned settlement (occupied for 5,000 years), a gentle dip in the coast hosts ancient, ankle-high outlines of a tiny chapel. Grand views sweep out across imposing skerries, while just one rough ridge away are tropic-like lagoons split by stoneworks: fish traps and lobster stores. Once it does appear in texts, at the end of the fifteenth century, Pabbay Mor is recorded as a place of natural wealth: 'ane fruitfull and fertile mayne ile, full of corne and scheipe'.[28] Modern soil surveys of another Pabbay, in the Sound of Harris, note 'heavily amended cultural soils . . . buried beneath wind-blown sand': this was, long ago, an agricultural centre where human action rendered soils productive.[29] At the Pabbay I now camped on, the ecclesiastical site – a vanished hermitage – sat on slopes boasting lush herbiage: a wide green oasis from exposed crags all around.

There is powerful evidence of rich, successful agriculture at Pabbay sites before and during the Viking age; not one such spot would be a natural retreat from the world of trade and travellers. It seems likely the Norse bothered to give these places names because they used them. One recent research project proposed that Vikings relied for provisions on the network of sites they called Pabbay.[30] In Ireland these seafarers drew on monasteries for grain and 'cattle tribute', taking only limited offerings so as to leave these places thriving. The Pabbay sites, it seems, might have been a northern equivalent: organised ecclesiastic farms that enriched rich land and ran surpluses from which armed voyagers could skim tribute.

As I travelled I'd seen that Pabbay sites were never really remote nor ocean-bound: they sat strategically at mouths of lochs and sounds. My final Pabbay, south of Barra, stood where routes down the island chain's east and west met before splitting again for Ireland or Argyll. Pabbays were, it seems, ninth-century service stations: stepping stones to fuel a voyage through Norse and Celtic worlds. As archaeologists since the 1980s have increasingly recognised, observing from the sea can transform perspective. Rethinking landlocked preconceptions has again driven home the rich geography and cultural significance of islands whose histories were once consistently undervalued. But the abandonment of small islands is not as easy to turn around as the minds of writers, and it seems unlikely that any isle named Pabbay will ever be inhabited again.

SUTHERLAND AND ASSYNT
(November)

A ship sails clean out of its metaphor
And birds perch on no simile; and Time
Breaks all the rules of reason and of rhyme.

Norman MacCaig

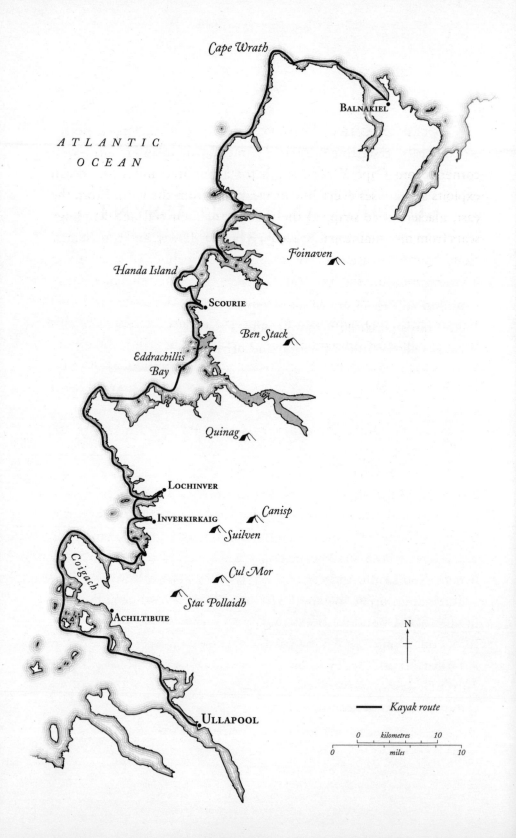

Cape Wrath

BALNAKIEL

ATLANTIC
OCEAN

Foinaven

Handa Island

SCOURIE

Ben Stack

Eddrachillis
Bay

Quinag

LOCHINVER

Canisp

INVERKIRKAIG

Suilven

Cul Mor

Coigach

Stac Pollaidh

ACHILTIBUIE

N

—— *Kayak route*

ULLAPOOL

0   kilometres   10

0   miles   10

AFTER LEAVING PABBAY, my journey leapt back northwards, restarting Scotland from the mainland's north-west corner. From Cape Wrath, south for a hundred miles, the ocean exploits and abuses every line of weakness from the west. From the east, glaciers once stripped the uplands of their soil and left huge scars from mountaintop to sea loch. Relentlessly wet, and consistently cloud-bound (occasionally recording under an hour of sunshine in a whole month), the resulting landscape is what the pioneering naturalist Frank Fraser Darling called 'largely devastated terrain'.[1] Few machair fragments survive among the bare headlands of this region. The most substantial and delightful, a dazzling contrast to the vast lithic bleakness, is Balnakiel: a broad bay beside Cape Wrath, watched over by a hulk of a farmhouse and a tiny ruined chapel (figure 5.2).

I began November's journey from the graveyard at Balnakiel. Light snow whitened the tops of the leather-brown hillsides, while the sea glowed deep blue except where wind scattered it in frothy blizzards. Hints of green persisted in the bent grass of the dunes, and occasional whooper swans and divers, always alone, flew over the bay's sheltered waters. Before I set out I found a small crude gravestone whose shallow inscription was upholstered with thick lichen: 'Rob Donn, 1777'.

It felt appropriate to gain the freedom of this coastline at the time of year Rob Donn loved best. As a young poet in the 1720s, Donn (pronounced like 'Down' by Gaels) was forced to spend summer as a farmhand in the byre of the tacksman, his master. Only in November, once harvest tasks were done, was he freed to lead his cattle high into the coastal hills:

But if autumn would come
I would not be wielding the flail.
There would be no concern about commerce,
Even if there weren't a bannock in the land.
There would not be a step between the Grey Hill
And the waterfall in Dougall's Corrie
That I would not be at liberty to follow.[2]

Over the centuries, countless Highland labourers have celebrated the onset of autumn leisure in similar terms, revelling in the end of the thousand tasks and blazing light of summer, when 'one may lave one's eyes with duns and russets and olives and golds and greys'.[3]

My journey this month would take me from a region sung in a hundred love songs in Rob Donn's Gaelic, to a place linked to a landscape poet from two centuries later. Norman MacCaig's output is defined by the coasts of Assynt but also by a sense of collective and personal loss. He spoke only English, which left him unable to converse with his maternal grandmother or Aunt Julia, both Hebridean Gaels; he possessed a fluency in the currencies of the Edinburgh public school – Latin and Greek – that he lacked in the language of his own Highland family. His sharp- and bright-eyed poems, in their small scale, delicacy and use of coastal landscape as the setting for relationships (whether human, human and animal, or even human and boat) echo Donn, despite their transposition into entirely different idioms.

Paddling down the deeply indented coast from Cape Wrath to Coigach, with poetry as both real and imaginative baggage, felt like a journey from an old Gaelic galaxy into newer English-speaking constellations. But ending at Ullapool – as the port to Lewis, constantly alive with Gaelic, where new Gaelic traditions are sung at the ever-bustling Ceilidh Place – brought me to a place with a richer cultural outlook than MacCaig foresaw.

Rob Donn's realm is perhaps the most elegant landscape in all Scotland. It's a rough but spacious plain of irregular pools and low rock ridges. From this wide 'knock and lochan' base rise small, bare mountains, rarely caked but often dusted in snow. And each statuesque hill is proudly independent: these are not the sociable hills that make up a mountain range. In their solitude, they're intricate and shapely (figure 5.1). Since their charisma is never compromised by the grasping shadows of other tops they glow at both sunset and dawn. Yet these hills are deceptively complex. During a walk up their slopes they slowly unfold their sweeping ridges, deep corries, sharp spurs and tumbling cascades of glacial clutter. This makes them a great adventure to sleep on between days of exploration. A still winter's dawn on Ben Stack or Ben Loyal brings with it a sense of vast, preternatural silence in which every falling rock or sliding snow patch resounds across the landscape as a grand event.

Over the years, I seem only to have visited in winter: I've never climbed my favourite hill here, Foinaven, except in crisp snow or showers of hail. On one visit the white southern slopes were lined with wild-cat prints and the frozen night air was ripped, for hours, by the weird rasps and snarls of a retching fox. Blizzards then set in, and at dawn I found myself in a gut-wrenching predicament. This was a lone descent of the vertical north-west edge of Foinaven's neighbour, Arkle, in a white-out when my crampons struggled to bite the colossal wall of snow and ice whose shapes below were hidden by the fog. My legs turned to limp spaghetti, my hands to brittle ice. I remember reassuring myself as I made each precarious step that I'd never again attempt something so stupid (but that, it seems, is a conviction all the best ocean and mountain memories have in common). The following year, I wandered up this way at the end of autumn, feeling like an awful human being. I'd abandoned a sick friend – the historian of Syria, Ben White – to shiver in a draughty bothy a few miles south: after driving up from the English Midlands, I wasn't humane or grown-up enough to sacrifice my fix of Sutherland's coastal mountains.

But I'd challenge anyone to exercise self-control when faced with these perfect hills. The views from the tops are achingly beautiful. Where the ground falls away beneath the feet, its shapes are like Euclidian art. Parabolas, hyperbolas and hyperbolic paraboloids unwind to the watery world below. Mathematical forms that are rarely found in landscape abound. Arkle's strange, flat pavements of hexagonal quartzite, 700 metres above the sea, feel like stylised landing strips, suspended in airy isolation. The rough dark plains sweep below, and it's immediately evident why land travel is slow and inefficient. Lochans, often ranked in parallel, can be miles long but only metres wide. They're shards of mirror to the red light of sunset: less like landscape than abstract art. Messy ridges of addled rock, ranged between the shards, resemble the bubbling surface of a boiling black broth, making the whole scene seem to exist in a different geohistory from the normal world beyond. In the winter sun, these immense views can be subject to ravishing light, thanks to clear air so far from anything urban. Whether under brief sunburst or high storm cloud, the flat world laid out beneath the steep slopes makes taking off in soaring, birdlike flight feel not just plausible but necessary. No wonder so much Gaelic verse takes owl-eye or eagle-eye perspectives on the world.

But these are hills of remembrance, not careless ecstasy, and the world at their feet is not as it was when Rob Donn knew it. Here, more than anywhere else in Scotland, the viewer is struck cold by a wholly modern emptiness. This is part of Sutherland, an expansive county with the lowest population density in Britain. There are just two people per square kilometre; in England, on average, that figure is nearly 500; in some British cities it is over 1,500. The modern population of Sutherland is half what it was in the early nineteenth century.

The population wasn't just reduced at that time, but also redistributed: those who remained here were pushed from productive inland pastures towards the coast. In this region, as in many other sites of clearance, shoreline sustenance was associated with poverty. Old Gaelic songs make diet a marker of class. The high-born girl, for instance, asserts her status: 'our homes are very different at sunset – in your father's house there are piles of fish bones, in my father's home there are venison haunches'.[4] The hardships and injustices of that period are such that when the poet Sorley MacLean's brother, Calum, first visited Sutherland he wrote that looking on these hills is enough to 'make any thinking Highlander feel murderous'.[5] It requires little alertness from the traveller through large unpopulated tracts, such as Cape Wrath itself, to find evidence in stone and scarred land of the shielings, crofts and field systems from which people were gradually, violently, extirpated.

It's difficult to know how best to respond to a landscape run through with such stories of brutality as are recorded in Jim Hunter's history of the Sutherland clearances, *Set Adrift Upon the World* (2015). It's clear that the nature of the destruction, carried out by people like George Granville Leveson-Gower, marquis of Stafford, needs to be remembered. But still more important is the memory of what they destroyed: the fame of Rob Donn should match the infamy of Stafford.

Born in 1714, Donn lived in a time and place when song was history and history song. Spoken and sung verse could be spread as fast as were, elsewhere, the books of the new lending libraries. Throughout the British Isles, in the era of Alexander Pope, poetry played profound social roles that it has since lost. But in the culture of Gaeldom the versifier's function was still more indispensable. Performed poetry was the medium of politics, humour, news, fun and faith; it was the vehicle of memory among a people who rarely consolidated their knowledge and experience in text. This helps explain why, to this day, poets are more crucial to comprehending the western corners of the British and Irish isles than to any other regions of the land mass.

Rob Donn's poetry is among the richest historical sources on life in the far north-west. Strangely unknown beyond Gaelic culture he and his two great contemporaries – Duncan Ban Macintyre, from Glen Orchy, and Alasdair mac Mhaighstir Alasdair, tutor to the 'bonnie prince' Charles Stuart – are renowned among Gaels. All three were born into an unsettled world. The conversion of the country from Catholicism to a strict Calvinism was newly, and incompletely, achieved, and in 1707 the English forced a Union on Scotland that was as violent and unwelcome as that of 1801 would be for Ireland. For every Scot in favour of Union, wrote Daniel Defoe (then an English spy in Edinburgh), ninety-nine were against. The ancient Caledonian parliament was dissolved, beginning a dark age of almost three centuries when Scotland's democratic apparatus dwelt in London. The parliamentary dissolution was lamented as 'the end of an auld song' while the bells of the national church of Scotland rang out a familiar melody: 'Why Should I Be So Sad On My Wedding Day?' Riots in Scottish towns were forcibly quelled. Far less surprising than twenty-first-century enthusiasm for independence is the fact such a manipulative 'Union' lasted its first few decades.

After 1707, the repercussions of shifts in religion and identity played out as conflicts between rival royals. The Protestant Hanoverians aimed to consolidate their grip on the throne against the continued claims of the Catholic house of Stuart. Although the Stuart (Jacobite) cause was intimately associated with Highland Scotland, Balnakiel was deep in the country of the Sutherland

Mackays; as such it was an island of Hanoverian sympathy in a vast Stuart sea. The diaries of the local clergyman record the laird's impatience to travel south and support his monarch, but they also reveal persistent frustration at the month-long lag between events in the Lowlands and the arrival of news in Mackay heartlands. While other regions of the Union became increasingly integrated through new arteries of commerce and war, this region of rough seas and rougher land began to feel its remoteness keenly.

Rob Donn, then, was on the side of Hanover but refused to use his verse for politics. In 1745 – the most resonant year in all Scottish history – the last great army of the Jacobite cause swept through the lowlands. But when they were decisively defeated at Culloden, an unprecedented humiliation was imposed on the Highlands: King George punished all Highlanders irrespective of whether they fought for or against him. This was the culmination of an era one Irish poet, Dáibhí Ó Bruadair, described in the imagery of an ocean apocalypse. '*Tonnbhriseadh an tseanghnàthaimh*', or the 'wave breaking', denoted the final cataclysm of an old order, when Irish and Scottish Gaeldoms were undercut and overrun by Anglo culture.[6] The new king's agent, 'Butcher' Cumberland, imposed a reign of terror across the north.

Donn snapped. Suddenly he poured out searing, Stuart politics. In screeds of treasonable verse he called on Prince Charlie to return, 'set the country alight' and 'avenge the day of Culloden'. He instructed his countrymen to 'have your teeth ready / Before your mouths are muzzled'. That Donn, summoned to stand trial in Tongue, got away with a caution for his treason (and then composed verse mocking the 'evil' court that tried him) is testament to the wit and sophistication with which the poet defended himself on the stand: he extemporised eloquent hymns in praise of King George and claimed these had been hidden from the court by their mutual enemies.

Politics, however, is not what Rob Donn is known for. Pale, dark-haired and brown-eyed, 'no taller', in his words, 'than a kid' but (in the words of others) rather stouter, Donn was an illiterate peasant. Two of his contemporaries passed judgement on his appearance; one called him 'good-looking', the other 'by no means a good-looking man'. Donn spoke hardly a word of the tongue of his nation's capital and lived entirely within the ancient Gaelic oral tradition. In the words of his biographer, his verse became 'the last and most complete picture of ancient Gaelic society that survives'.[7] Donn was no John Clare, lamenting the loss of a way of life, nor Wordsworth, singing the praise of a wild landscape, but a poet of a thriving culture, who satirised, praised and lampooned the social world around him. He gave detailed portraits of the weaver, the parish priest and the families of the lairds. In a newspaperless society, he versified notable events, including the occasion when a naked sailor went running through the region seeking trousers and met with many kinds of ribald mockery.

The encounters Rob Donn valued most took place not among byres and chapels but on deer slopes or during nights outdoors when tending his herd. His youthful idea of paradise was a lone meeting with Isabel, his master's youngest daughter, among the black cattle on the slopes above the cape. In one poem, Isabel and her haughty sister converse. Isabel is full of love for the tasks of coastal field and mountain stream. Mary, schooled in the fettled English-speaking town of Thurso, far to the east, mocks the traditions of her family as quaint and rustic. In such poems, Donn elaborates the distinctions between the two cultures that were competing for the Highlands' coastal heart.

Donn's verse portrays landscape entirely differently from the English traditions of his era. It is striking that the hillsides are both his workplace and his escape. Wishing to be subjected to no authority (a role in which he later casts his apparently long-suffering wife), work on the hillside is dignified and life-affirming in contrast to the drudgery of indoor effort. Elsewhere, attitudes to landscape were largely utilitarian, so that to writers such as Samuel Johnson, the hills where Donn found delight were merely grim 'matter incapable of form or usefulness'. English-language perceptions of landscape changed across the eighteenth century: by 1790, in the hands of Romantic poets, they resembled Rob Donn's enthusiasm of 1730.

What separates Donn from the Romantics is that when he wrote of awe or joy in response to land, sea or weather, these were never the subject of his verse, but the context for events. He describes, for instance, a change in the weather while a party of Macleod men embarked south, along the coastline I'd soon kayak, before they crossed the Minch to Stornoway:

> In the morning we were obliged,
> When the wind rose to a gale,
> To turn our backs to the land
> And our faces directly towards the sea,
> Subjected to the drenchings and the beatings
> Of the furious great waves
> Mountainous, foamy, stormy, deep-valleyed,
> Sucking, thick-lipped, blue.

But those seas are merely a stage Donn sets for the actions of his comrades:

> As she made headway
> Forwards on her course,
> The Macleods were unerring and expert
> About the sheets of the sails.
> Watchful, mindful, powerful
> Was Patrick at the helm,
> And George Roy of Tarbert was there
> Doing the work of three.

That crew alternated
With fortitude and fear.
They were to be seen smiling
Though they had left rock and mountain behind,
With unfainting courage,
Without timidity in their actions
Though not a man of the five of them
Had ever set foot in Lewis.

But the divine helmsman looked mercifully
On our plight in time,
When it was impossible to boast
Of sailor or of carpenter.
From the back of the great-troughed sea
And the water rushing forward in a torrent
And from the summit of the whirling waves
She struck her prow against land.

In Stornoway, the poet takes up the themes of Lewisian hospitality and the popularity with women of George of Tarbert, Lord of Handa Island ('Goodness is he married?'). The journey, like the encounters in the town, are Donn's means for establishing the character of his true subject: this is, in fact, a praise poem written for George, who paid him a horse for his trouble.

I travelled westward from Donn's burial place, launching through surf and plunging into a still morning's high and 'thick-lipped' swell. The first two days of this journey took me along coastlines that eighteenth-century travellers sought to avoid. Donn wrote of a man identified only by first name (which happened to be the same as mine) who somehow survived being swept to Orkney while attempting to round the cape.

How distressing and miserable the crossing
That David made to Orkney –
There went the cheese, the kelp and himself.
Your neighbours were on the lookout,
Seeking news in every bay,
And such was their anxiety that they couldn't shed a tear.
But when they heard of your return
From the seas without mishap
Then people shed tears aplenty.

I thought fleetingly of the frayed nerves of my own family as I set out on these restless seas; but my mind was dominated by musings on the significance of poets like Donn to the Gaelic culture I'd experienced in the Western Isles. If English was confined to a small pocket of Cumbria or East Anglia, pushed up against the sea by a neighbouring culture that couldn't see its worth, then its literary tradition would surely be a rallying cry for its defenders: the desire to comprehend the subtleties of Shakespeare would be a reason why the language must not be deprived of the fluency that only everyday use can bring. In a similar way, the poetic inheritance inspires the confidence with which Gaels today sing the virtues of their culture in ways to which their grandparents were less inclined. As the playwright Norman MacDonald put it, the world has begun to swing Gaelic's way:

Everywhere you look there is a new emphasis on the importance of the small community. There's a new determination to work with nature, not against it. There's a growing disenchantment with the huge, unwieldy, thoughtless, feelingless cultures that have dominated the world for the last two hundred years . . . Gaelic will be swept along on that tide . . . Our poets alone have made that essential. The reservoir of poetry that we have in Gaelic is so important to humanity that we just can't let the language die.

The work of these Gaelic poets (like the Counter-Desecration Phrasebook) has the particular virtue of being tied to place: landscape and language are resources that constantly enrich one another.

The cape itself is an entirely other-worldly cliffscape: as remote as anything in Britain yet still as richly inscribed with names. The long westward approach passes a series of huge offshore skerries. Each is scarred, bare and brutal, but that is where their resemblance ends. The region is like a vast gallery of modern art with each exhibit sculpted by a different artist and named in Norse or Gaelic. An Garbh-eilean is a high wall of solid rock, simple and monolithic but cratered like a moon of Jupiter or Saturn. Its name is Gaelic for 'the Rough Isle' and is echoed in Garvaghy in Ulster. Next comes Stac Clò Kearvaig, a tall tower bristling with spikes, like some airy hill-fort; its name derives from a Nordic ship, a little smaller than a longboat, and echoes Karvevik on the north Norwegian coast. Thousands of miles of Atlantic seaboard are thus knitted into a few ragged chunks of rock that have never been farmed, never desired, rarely, if ever, even stepped on. By the time I reached this point, the blue had turned grey and rain had set in. I began to plunge through tidal overfalls and the grandeur of the shadowy stacks showed more forbidding aspects.

After a long and bumpy ride across the swell-strafed mesh of tides that gives Cape Wrath its name (derived from the Norse for turning point, *hrof*) I was on the southward journey that George of Tarbert paid Donn to sing. For two days I delved into the sheltered inlets of Loch Inchard and Loch Laxford, looking up at the low slopes of Foinaven and Arkle. Despite leaving Donn's land behind, there was still something wistful about this landscape on these short autumnal days. Scots pines creaked above, while brown bracken and heather decayed beneath them. I sat to rest by a small stream and looked up to see a tiny deer calf trot towards me. The most beautiful of creatures, it was large-eared, with deep black eyes and small wet muzzle.

The fur on head and neck were silver and short, but its back was a thick russet fuzz, flecked with black, that seemed to demand a hand run through it. The calf stopped an arm's length away and stared as though expecting a response (figure 5.3). It was entirely alone: there was next to no chance it would survive the coming winter.

The next leg of November's journey brought reminders of connections by land and water. On the coldest morning of the journey so far, with slow swell but a frenzied chop, I made my way to Handa Island. Passing beneath the island's towering sandstone cliffs on a rough sea was as challenging as Cape Wrath itself; in the seam of tidal movement at the north-west corner, I found myself swimming: tipped from my boat and kicking free after failing to roll for the first time in months. Half an hour later, my teeth chattered as I watched adult sea eagles feed their giant offspring in the heather.

Home in summer to 200,000 seabirds, Handa is now the premier bird reserve of the far north-west. Run by the Scottish Wildlife Trust it feels, in winter, as wild and unpeopled as anywhere on these coastlines. Yet its historical significance belies its littleness. Throughout tales of the northern clans, Handa recurs as a place whose people changed the course of history. I'd first come across the island's reputation as the home of large and powerful fighting men when reading tales of the sixteenth-century clans of Atlantic Lewis. In 1595, John Morrison, brieve of Lewis, conspired in the murder of the Macleod chief, Torquil Dubh. The Ness Morrisons fled to the mainland to escape revenge. But a broad-shouldered Macleod warrior, John mac Dhomhnuill mhic Ùisdein, lived on Handa and it was he, with a small band of men, who punished the Morrisons, despite their greater numbers. When other men of Ness crossed the Minch to retrieve the brieve's body, a storm prevented them returning home and in order for his corpse not to decompose entirely they were forced to bury his spilled and rotting innards on an islet of Eddrachillis Bay (just south of Handa) still known as Eilean a' Bhritheimh: the Brieve's Island.

At that time, Handa itself was a place for burials. The wolf had been hunted close to extinction, but the tradition of using islands to save corpses from wolf packs had not yet ended. Later collections of verse record wolfish disturbances:

> On Eddrachillis' shore
> The grey wolf lies in wait, –
> Woe to the broken door,
> Woe to the loosened gate,
> And the groping wretch whom sleety fogs
> On the trackless moor belate.
>
> Thus every grave we dug
> The hungry wolf uptore,
> And every morn the sod
> Was strewn with bones and gore:
> Our mother-earth had denied us rest
> On Ederachillis' shore.

South-west Handa, where the ruins of a chapel can still be seen, hosted multitudes of burials, but the rest of the island was populated far more heavily than much of Sutherland; evictions took place in 1829, but only in the 1840s did the last of the population (sixty-five in 1841) quit Handa. Like St Kilda, the island was run by its own small parliament where, at daily meetings, the oldest widow was queen. Yet documentation of Handa traditions is far sparser than that for similarly historic centres: it would take a novelist rather than a historian to restore the queens of Handa to life. Here, as at Havera on Shetland, the dispersal of island people was followed by avian colonisation. No skuas bred here as late as 1960, now 250 pairs dominate the interior: one pair for each human resident in 1800.

Bearing down on Handa is Ben Stack: a small mountain of steeply pyramidal form that marks the northern boundary of Norman MacCaig country. The improbable act of imagining Handa densely peopled has a parallel in MacCaig's verse. Standing on Ben Stack one day he recorded the strange historical dislocation of seeing old sea roads resurrected: a 'real full-rigged ship in this wilderness of a place'. To watch a vessel of the old clipper days pass Handa was, he said, 'like something from another time':

> At the cairn I turn round and scan
> the jumbled wilderness
> of mountains and bogs and lochs,
> South, East, North and then – West
> – the sea
>
> Where a myth in full rig,
> a great sailing ship, escaped
> from the biggest bottle in the world,
> glides grandly through the rustling water.[8]

Time and the sea are twin obsessions in MacCaig's verse, and to a Scot of his temperament each was both beautiful and disastrous. 'History frightens me', he wrote, its developments bringing 'the goodies of civilisation, / every one sweet, every one poisonous'.[9] The fate of the Highlands under the commercial pressures since Donn's time meant that MacCaig was deeply sceptical of his era, prepared only to 'put one foot / dangerously into the twentieth century'. No nationalist, he famously stated that

My only country
is six feet high
and whether I love it or not
I'll die
for its independence.[10]

Yet he was more attentive to the damage nationalism did to local difference than the ways in which it toxified international affairs. Despising abstractions, he saw 'patriotism' as an emotionally laden 'big word' – along with progress, liberty 'and all their dreary clan' – each used in unexamined ways to trigger stock responses from half-attentive listeners.[11] Poetry was a tool for encouraging scepticism towards the 'fake, the inflated, the imprecise and the dishonest' and for seeing and celebrating things divorced from the pompous and the 'high falutin". Although he had an intense love for individual people – people with unique personalities and stories – MacCaig insisted that he hated 'humanity' in the abstract: humanity, after all, was the species that killed on industrial scale and desecrated land-scapes with equal abandon. This attitude, born of war, was shared by many Scottish poets of the era. Hugh MacDiarmid, for instance, had written that all the 'big words . . . died in the First World War'. But it was also the response of someone attuned to his environment at seeing the scale of man-made degradation: it's now the explored areas of the map, MacCaig famously wrote, in which people write 'here live monsters'.[12]

MacCaig's view of landscape might be seen as similar to his percep-tions of humans and humanity: he isn't the kind of poet to praise a mountain range or a long river; instead he conjured the detail of a specific spot and the precise thoughts triggered by its observation. It was at sites on the coastline south from Handa where he honed these skills of seeing. Time and again, observing the detail of landscape, or the movement of sea, inspired a characteristic stretching or compression of time. In the poem 'Wreck', for instance, he considers a stranded hulk:

> Twice every day it took aboard
> A cargo of tide; its crew
> Flitted with fins. And sand explored
> Whatever cranny it came to.
>
> Its voyages would not let it be.
> More slow than glacier it sailed
> into the bottom of the sea.[13]

In his most famous and anthologised poem a personal but primal encounter with a 'room-sized monster with a matchbox brain', the basking shark, leads him to muse on the primeval monster in himself. Rowing in a bay thirty miles south of Handa, the accidental impact of his oar on the unseen shark is the moment when this 'decadent townee' shakes 'on the wrong branch of his family tree'.[14] In another poem, MacCaig confronts a Neanderthal and sees in him 'what civilisation has failed to destroy in me'.[15]

More perhaps than any other poet, MacCaig was profoundly attuned to the ways in which engagement with the sea conjured the early stages of a slow, evolutionary and historical detachment from the water. Boats to him were the stuff of early epics – from Homer's *Odyssey* to its parallel in Gaelic, Alasdair mac Mhaigstir Alasdair's *The Birlinn of Clanranald* – which were reprised in miniature every time a person embarked on a sea voyage.

The 'family tree' that mattered to MacCaig was his Gaelic heritage. Seeing himself as a 'leaf that hangs down helpless' from that past, he bemoaned his split condition – a Lowland English-speaker with island Gaelic roots – and envied, achingly, those who could be at one with past and place (it's hard for the sandpiper, holds one saying of these dual-cultured coasts, to work both the ebb tides). These convictions were tied intimately to his love of wilderness: 'comfortless places comfort me'; they were also tied to the ways of life – crofting, fishing, weaving – that once took place in all the ruined structures of the land he loved.

Yet MacCaig's most fiercely emotive poetry is in fact urban, recounting his arrival in the town of Ullapool during a deep northern freeze. He drove through thick, unblemished snow – not that 'horrible marzipan in the streets of Edinburgh' – and entered a bar 'fireflied with whisky glasses' to meet his great friend, Angus MacLeod.[16] He likened the two of them to an accordion and fiddle that 'fit nimbly together their different natures'. But this was the winter Angus died. Through his grief, MacCaig felt that the death of the man with whom he'd explored loch and hillside made 'the pagan landscape sacred in a new way'. The funeral took place beside the 'boring, beautiful sea' and, later, MacCaig's favourite themes poured out as he described music, in an Ullapool bar, that bounced with such wit that the darkness felt small:

Out there are the dregs of history. Out there
Mindlessness lashes the sea against the sea-wall:
And a bird flies screaming over the roof.

We laugh and we sing, but we all know we're thinking
Of the one who isn't here.

At the opposite end of coastal MacCaig country from Ben Stack, Ullapool is the big city and bright lights of the far north-west and was the culmination of this month's journey. With 1,400 people it was the first town of more than 1,000 I'd passed for months. On my travel so far, only Ullapool and the Orcadian town of Stromness (accessible from the Atlantic but facing east to the North

Sea) had a population of over 600. The recent past of Ullapool's single mile of shore thus reveals elements of coastal life – from large industrial shipping in the 1970s to the biggest drugs bust in UK history – at odds with the stories raised by the other 600 miles I'd now travelled.

Designed by Thomas Telford and built by the 'British Society for Extending the Fisheries and Improving the Sea Coasts of this Kingdom of Great Britain', Ullapool didn't begin as a cogently Gaelic community but as an imposition. Through decades of gradual integration with neighbouring villages, it retained a bilingual existence: where most places nearby were over 90 per cent Gaelic-speaking, Ullapool peaked at around 75 per cent, and an unusually high proportion of those speakers could read English but not Gaelic. There were few residents who spoke no English at all: around 10 per cent compared to 50 per cent a few miles north. In the 1830s the Statistical Account noted that Gaelic was losing ground in the whole region: Ullapool was a conduit of expanding Englishness. By the time MacCaig visited, few words of the language would be heard on the streets unless among tourists from the Western Isles.

MacCaig died in the 1990s just as the developments that were rejuvenating Gaelic in the islands began to be felt in his mainland haunts. And the most surprising feature of this development was that urban centres like Ullapool, which had led anglicisation, now drove the cause of Gaelic education and culture. A Gaelic primary school was founded in the town in 1991; many of the children came from backgrounds – both local and incoming – in which Gaelic wasn't the main medium of life. Cities as far outside the Gàidhealtachd as Glasgow and Edinburgh have also led the urban Gaelic renaissance that followed hot on the heels of island revival. Where Gaelic when MacCaig died was the heritage of the rural elderly, today it's also the future of the urban young.

In 1970, the actor Robert Urquhart opened a small room behind

Ullapool harbour which invited musicians to perform for their supper and share their histories. Now, that room has expanded to be the town's leading venue: the Ceilidh Place. This bar increasingly caters to young Gaels, with a festival of Gaelic culture each September, Gaelic books in its bookshop, and Gaelic singers through the year. If MacCaig could return to a fireflied Ullapool bar once more, the verse his visit would inspire might well express more optimism than the regretful stance he's known for.

A MOUNTAIN PASSAGE
(December)

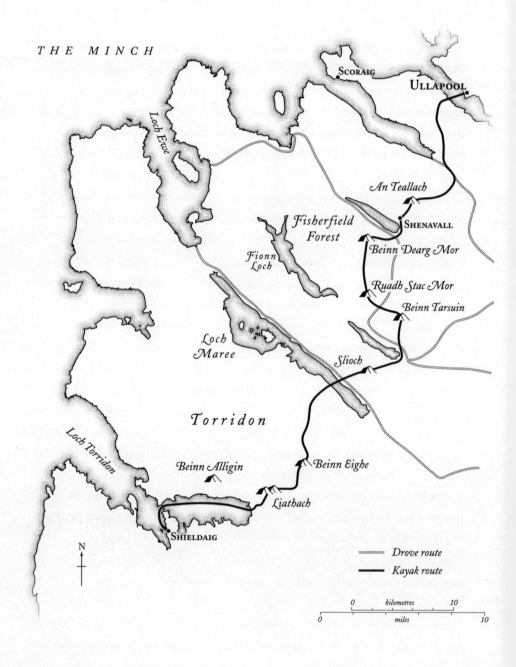

THE MINCH

SCORAIG

ULLAPOOL

Loch Ewe

An Teallach

Fisherfield
Forest

SHENAVALL

Fionn
Loch

Beinn Dearg Mor

Ruadh Stac Mor

Beinn Tarsuin

Loch
Maree

Slioch

Torridon

Loch Torridon

Beinn Alligin

Beinn Eighe

Liathach

SHIELDAIG

N

Drove route

Kayak route

0   kilometres   10

0   miles   10

DECEMBER, AS USUAL, promised storms. I had little chance of kayaking far without long and miserable confinement to spots where weather made movement impossible (figure 6.1). So I abandoned the kayak and headed uphill, stuffing my rucksack with the tiny packraft that would carry me over three great lochs as I made my amphibious way between Ullapool and Shieldaig. Mountains and the western seaboard are synonymous, as are hills and the Gàidhealtachd. *Anail a' Ghàidheil, air a' mhullach*, runs one saying: 'the soul of the Gael is on the summit'. The sea routes of the Minch and the Atlantic were so significant precisely because the frictive forces of uplands rendered land travel gruelling and costly. Today, the population of this region is almost entirely concentrated on the coastline and its eastward skylines are always mountain.

Only the Cairngorm range rebuts the rule that the highest hills rise from western shores; with the Atlantic influence attenuated, everything of Cairngorm – structure, climate and species – diverges from the rest of Britain's uplands. In contrast, the ranges south from Ullapool are perhaps the most emphatically oceanic, embodying the ragged ridges over deep sea lochs that indicate the twin erosive forces of ice and sea. It's an undulating land of a hundred tight-packed hills, with zones whose names – Dundonnell, Fisherfield, Torridon – draw panegyric from countless climbers, walkers, geologists and geographers. A winter shell of snow and ice makes these peaks a place of real adventure: they are, like all Britain's mountains, relatively small, but they inspire far more writing than ranges many times their height. Indeed, the west of Scotland is among the best-documented highland on the planet.

Only a tiny proportion of such writing features the history of those who worked the land, because these hills have never generated extensive historical record, and the sites that archaeologists would otherwise dig lie far from modern points of access. These familiar hills thus do little to shape accounts of British history or ideas about the ways of life that were widespread in these islands' past.

That's not to say there haven't been fits of mountain enthusiasm among historians. Most, regretfully, make cautionary tales, not inspiration for a new upland perspective on the past. Such cautionary notes fill my memories of early interest in mountains and history. When, in my early teens, I found a volume by the historian G. M. Trevelyan that is mentioned at the beginning of this book, the first two things I read were brief but incandescent essays. The first, 'Clio: A Muse', was an outpouring of anger, penned when the young Trevelyan sat through a lecture by the most venerable professor of history in Britain, J. B. Bury. Bury declared that the discipline of history must be conducted on scientific lines. Trevelyan lamented that, under the auspices of such anodyne ideals, 'literature, emotion and speculative thought' would be banished from how people interpreted and expressed the meanings of the past.

The second essay was simply called 'Walking' and in it Trevelyan explored the place of mountains and glens in his vision of what it meant to be a historian. His romantic belief in the roles of wild places in historical thought was infectious, if judgemental. He called his legs his 'two doctors' and wandered upwards of fifty miles a day upon them. Not all walks, he pronounced, had equal recuperative power: 'you cannot do much with your immortal soul', he insisted, 'in a day's walk in Surrey'. Instead, he extolled the northern torrent of molten peat-hag that we ford up to the waist, to scramble, glowing warm-cold, up the farther foxglove bank; the autumnal dew on the bracken and the blue straight smoke of the cottage in the still glen at dawn; the rush down the mountainside, hair flying, stones and grouse rising at our feet; and at the bottom the plunge into the pool below the waterfall.[1]

The experience of being in wild places, Trevelyan insisted, was full of insights into past ways of life. Finding mountain routes in the mist was, for instance, 'one of the great primeval games'.

In my teens, I was instantly on Trevelyan's side in every debate about history. It was only later that I learned more of the context to his ecstatic scree-slope screeds. Only slowly was I sensitised to the things that evaded his aristocratic gaze. Hurtling down the hillside, enjoying, because of his status, a freedom of the moors denied to most, Trevelyan would burst into the small Highland croft whose smoke he'd sniffed from the ridge and demand food and shelter. He loved the people in these homes with the fiercest condescension, praising their 'honest cheerful poverty' and wishing that he too had a life so attuned with history and uncomplicated by wealth or power. His attempts to comprehend their lives never proceeded far: Highlanders were symbols of his à-la-mode struggles with urban malaise.

Throughout his writing on rural Scotland, Trevelyan aimed to show that the culture of motor cars and trains had not wiped out the wisdom and dignity of the past. His great-uncle, the venerable Victorian historian and politician Lord Macaulay, wrote that if the Britain of 1685 could 'be, by some magical process, set before our eyes, we should not know one landscape in a hundred, or one building in ten thousand'. Riled by this (he was, it seems, always driven by annoyance with something), Trevelyan set out to reveal the past's intrusions through the landscape's modern dross: 1685, he believed, could be met with on any wander through rough country. Imaginative connection with outdoor histories was the prescription imposed by his two perambulatory doctors: it sustained health because it revealed every absurdity of the city's speed and steel.

Trevelyan compiled a brief history of walking in which hillside escapes and escapades inspired critical, external perspective on self-absorbed cities. His walkers were visionaries and prophets: there were many schools of walking, he said, 'and none of them orthodox'. Though Coleridge and Wordsworth were the great instigators of 'walking into truth', particular praise was heaped on the Scottish historian, Thomas Carlyle – the 'seer of Ecclefechan' – presented as the scourge of instrumental rationality and all other world views that marginalised imagination. Carlyle walked large distances, gesticulating his way across the moor as he muttered purple prose. While wandering, he dreamed up vivid portraits of great events, expressed with unique intensity in a style soon christened 'Carlylese'. When walking, his companions wrote, Carlyle became a feature of the landscape: 'a living, not extinct volcano whose lava-torrents of fever frenzy enveloped all things'.[2] Neither he nor Trevelyan saw history or literature as a thing to be done at a desk in an urban room without something elemental to ignite the imagination.

Trevelyan became increasingly fond of phrases like 'sacred union with nature' and threw himself into causes such as the Pilgrims Trust and Outward Bound as a kind of crusade against urban life.[3] This was so central to his profile as a public historian that he was appointed president of the Youth Hostel Association on its inception in 1930. Press coverage of that moment treated the identities of historian and outdoors champion as self-evidently linked, never suggesting surprise that the two pursuits had somehow become enmeshed.

But subsequent decades, from the 1940s to the 60s, saw this attitude to the past perish, choked by professionalisation and the pursuit of scientific rigour. A few, such as the historian of the seventeenth

1.1 The waterproof sleeping bag that was my home for much of the journey. Ben More, Isle of Mull.

1.2 The vessel for the journey, an SKUK Explorer, designed for rough seas and long expeditions. Eynhallow, Orkney.

1.3 The picture that most sums up the reaction of wildlife to a kayak: mild surprise, and more curiosity than fear. This otter appeared and stopped, little more than an arm's length away. Great Bernera, Isle of Lewis.

1.4 Still days offshore mean encounters with species rarely seen from shore or larger boats. The little auk, a sparrow-sized relative of guillemots and razorbills, is among my favourite species to find on the water. Off the coast of Sutherland.

2.1 The view from the Hermaness Cliffs on Unst where I spent the first night, showing the waters I'd kayak next morning. The pyramidal form on the top right is the beginning of the Muckle Flugga skerries, with Out Stack just hidden behind the cliffs.

2.2 Looking up into the immense cloud of gannets that issued from Muckle Flugga. The sound of their wings drowned out the noise of the sea.

2.3 Rough sea practice with Llinos, before beginning the journey, from Hosta Beach on North Uist.

2.4 Kayaking through tides at the Ramna Stacks between Yell and Shetland Mainland: the first time I dared photograph such conditions while kayaking through them.

2.5 The extraordinary rock forms of Northmavine (Shetland Mainland). These sandstone stacks are often topped with nesting sites of rare seabirds such as arctic terns.

2.6 Nort Ham. Once the only safe
point of entry and embarkation
for the lost community of the isle
of Havera.

2.7 Fulmar and chick on Foula: one of
the few places without land predators,
so that cliff-nesting birds like fulmars
breed on flat ground. In July, the
north-east corner of the island is a
meadow of seabird nests.

2.8 The immense cliffs of Foula. Witnessing a storm here – with thunder
reflecting from the cliff face and thousands of seabirds wheeling out in
response – was one of the most elemental moments of the journey.

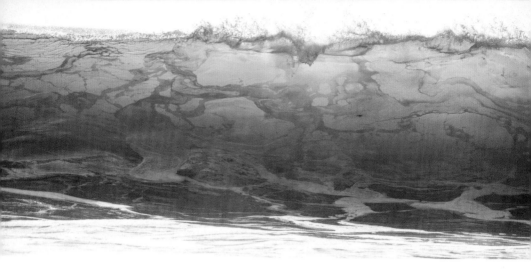

3.1 A kayak's-eye view of an oncoming wave.

3.2 The Holm of Aikerness (between the Orkney islands of Papa Westray and Westray): once a busy heartland of the industrial revolution.

3.3 A seal shooting station on Eynhallow (Orkney), evidence of the wealth that seal products could once bring islanders.

3.4 Challenging conditions round the Brough of Birsay (Orkney Mainland). With a rising swell, and failing light, this was a constant balancing act before a bruising landing.

3.5 One of the tiny harbour seals in the sheltered bay east of Eynhallow, surrounded by racing tides on either side.

3.6 The gannet that approached at Noup Head (Westray), clamping its bill round the bow of the kayak, before swimming alongside.

3.7 The west sides of Orkney and Shetland are famous for their sea-carved cliffs, caves reaching deep beneath the islands.

4.1 The dark land and the bright: lazy beds in the rich shell-sand ground of the Isle of Harris with the old hard rock and acid earth behind.

century, Veronica Wedgwood, fought to keep Trevelyan's poetics of outdoors history alive. She declared the landscape 'the hardest of all documents to read' since each century scrawled new text on 'its defenceless surface'. But in most of Britain the battle was, for a few decades at least, lost, and history retreated into urban topics and themes more in tune with Bury's drudging than Trevelyan's trudging. Bury's followers went still further than him, insisting that anything unsupported by statistics had no explanatory power: that which couldn't be counted didn't count.

Remnants of muddy-booted history survived longest in the Highlands, where another aristocrat of Trevelyan's ilk, Archibald Haldane, continued to wander the hillsides in search of the Scottish past. Born into wealth and educated in Edinburgh and Oxford, Haldane began his working life as a lawyer, spending spare hours exploring hill paths and streams with a fishing rod slung across his shoulder. His early writing recounted days of leisurely angling: *The Path by the Water* (1944) and *By Many Waters* (1946). But Haldane loved libraries almost as much as he loved trout. As chair of Edinburgh's city libraries committee and trustee of the National Library of Scotland, he found himself surrounded by archival records of the Highland routes and rivers he walked and angled. His interest was piqued and given peculiar richness by chance encounters, when out fishing, with old men who had once walked drove roads as they transported cattle from the Western Isles and northern Highlands south.

Haldane's books *The Drove Roads of Scotland* (1952) and *New Ways through the Glens* (1962) could not have been written by anyone without intimate walker's knowledge of the Highland paths alongside the detailed familiarity with Scottish community that Trevelyan entirely lacked. The results of Haldane's long tramps through wild glens thus retain significance when Trevelyan's have become vivid but uncomfortable museum pieces.

My route this month took a course like a wheel rim round drove-road spokes. All the regions I travelled in western Scotland – from Uist and Barra, to Sutherland, Assynt, Ullapool and Skye – had been headsprings of Haldane's cattle-droving routes. These regions would, at fixed points in annual cycles, be transformed by black longhorn livestock ambling through. The journeys began in the Minch in July, where ships gathered to seek employment in the transport of cattle from the Outer Hebrides to Skye. These ships were smacks: substantial sailing boats with a mast set back to permit large foresails which, although they reduced manoeuvrability, suited encumbered journeys through heavy weather. Their arrays of elegant red-brown sails made a scenic image of the sea lochs.

Few ports in the north-west could handle these cattle-filled smacks. When the boats reached shore livestock were often just pushed into the sea, salt water helping heal those gored by their neighbours in the crossing. Haldane describes the 'tryst' at Broadford on Skye where, at the height of summer, livestock were gathered for the next stage of their journeys. He conjures

> the noise of cattle and the shouting of the drovers, the quarrelling, bargaining and courting of the crowd, the tents in the hollow where food was cooking for weary drovers, and in the background the steep stony sides of [the mountain] looking down on the bay and the river.[4]

He describes the pristine reflections of shapely hills in sea lochs and wonders whether drovers ever saw this scenic workplace as compensation for difficult lives.

Cattle were swum from Skye to 'the continent' (as drovers called the mainland) tied tail-to-jaw in rows of six to eight. At slack tide they'd be led into their swim, the rope of the foremost cow gripped by a man in the stern of a boat while his colleagues rowed. There were no fixed routes these convoys took from the coast. Skilled drovers chose their paths according to weather and social circumstance (avoiding, for instance, sites of clan unrest). By October, 'thirty thousand black cattle in different droves overspread the whole adjacent country' where routes converged at Crieff and Falkirk. Each leg of the journey here would have carried risks and all drovers could expect total failure one year in ten, with insolvency their inevitable reward. But from the 1720s to the 1810s, when Britain's incessant wars inflated demand for salt beef to feed armed forces, a year or two of dearth was not too high a price for the many years of plenty.

Droving was an unusual lifeway because cattle were unique. Sheep were too small and stubborn to wade Highland rivers and were, along with inanimate goods, transported by sea road or inland waterway in preference to path or hill track, but cattle were biddable and sturdy. Wheels, required to transport goods like grain, would stick in mud and river, but cattle could swim and wade at least as well as drovers. Another factor to keep cows off the sea was their notorious misbehaviour on ships: countless island stories involve cattle overturning boats or stamping holes in hulls. Indeed, being stuck on a stormy sea with a bull was a nightmare worthy of legends: the occasions when that happened perhaps inspired the many coastal myths of demonic sea bulls that made this land-beast a symbol of watery horror.

Yet the response of droving to the nineteenth-century reordering of landscapes was exactly the opposite of other trades. This era brought roads and rail but also entailed new property rights, including the enclosure of land by fence and wall, which all but ended the practice of grazing livestock by the track and sleeping off-road among the herd. This agricultural improvement made drovers unwelcome on the back ways at the same time as tolls were introduced on major roads to fund the investment that rendered them passable by coach. Such expenses, accumulating over distance, created new costs that sapped drovers' profit.

So it was that in the era when many trades moved from the sea inland, the cattle trade left the land for the sea. Drovers had travelled light: cattle were fattened for market at their southern destination. But the development of larger sailing ships, and then of steam, allowed fattened cattle to be shipped. Commercial transport outfits could fund facilities for unloading fatted cows from steamers: market-ready stock began to be packed into ocean vessels bound for Glasgow and Liverpool or railheads like Stromeferry. The new practices weren't less risky than the old. The loss in condition of cattle at sea was great, and deaths occurred on every journey. Sometimes, in adverse weather, all would be thrown overboard or smothered in the hold. In storms, the terrified cattle did enormous damage. This was still a gambler's game. And there was always a long story behind every steak on a southern plate.

When north-western glens ceased to be cattle-ambled, the trades that took the place of droving were fitting for a commercial age; they took place in towns more than in open country. Those regions that were not enclosed for arable use were devoted to deer and grouse, and rail deposited shooting parties on increasingly sparsely populated land. The world of Trevelyan – where aristocrats waltzed through lands they saw as empty or innocent of commerce – was born.

After the age of droving the people who lived on this land were often stalkers or gamekeepers to the lairds, and their life, for a while, was spectacularly disconnected from the accelerating developments elsewhere. After two days crossing the peninsula of Scoraig and the peaks of the mighty mountain An Teallach I reached Fisherfield. This is a region famous for its lack of roads or buildings and therefore labelled 'the Great Wilderness', or the 'Whitbread Wilderness' after the industrial family whose estate this once was. Rivers here, though substantial, are unbridged and in winter must be waded, while most tracks that appear before a walker's feet were made by the pads of hare or hooves of deer. From the peaks at night, the distant lights of a few small settlements can be seen, but the blinking beams of far-flung lighthouses are the most striking human feature of the fabulous darkness. If remoteness was the reason I've been drawn so many times to these severe hillsides, the purpose of this journey was to diminish and dismantle it: to populate with figures from the past a landscape that I'd loved because of an illusion of emptiness.

In contrast to the storms that followed, the day I entered Fisherfield was bright, still and bitterly cold. As in the previous year, December snows hadn't materialised and without either veils of white or patches of summer green, large swathes of rock-strewn slope were bare, black and exceptionally bleak. Ice, cracked into shards like shattered glass, lay on every lochan and lined the sides of streams as it crept towards their centres. The gorsey riverbanks were the one place with more than a thin cover of vegetation. Each thorn was coated in hoar frost, but the deep gorse green had a deadening effect on the thick crystals, replacing their usual gleam with a flat and ghostly tone of blue. Once I'd descended into the wide river valley at Fisherfield's northern edge, I spent the night in one of Britain's most beautiful buildings: a tiny bothy wedged between the silent mountains at Shenavall (figure 6.2). Deer gathered round the door, and wrens flitted into holes in the mortarless walls as day faded into a frozen night of stunning constellations. I had nothing to build a fire, but lit three tea lights in the bare, wooden interior of the bothy, hoping vainly to gather a hint of warmth.

I'd slept here twice before, but each time I'd revelled in the wildness of the scene and paid little heed to the hillsides' human traces which, with the foliage stripped back by winter, were now uniquely legible. Nor had I learned much about the building and its early occupation, which I'd assumed to have been largely unrecorded. Fortunately, in 1980, a volunteer responsible for the bothy's upkeep, Alex Sutherland, had taken a ninety-four-year-old, Colin MacDonald, for a hill walk. Colin's childhood had been spent in this bothy, and almost all that's recorded of its early history was recounted to Sutherland on their stroll.

Shenavall bothy was first occupied on a morning at the start of winter 1891, when mist clung low round the mountains. This was at the height of the period when, as traditional land uses declined in the aftermath of clearances, many Highland regions were repurposed as sporting estates. In 1811 there were fewer than ten such estates; by 1891 there were 130, covering 2.5 million acres and owned by financiers, industrialists or aristocrats who used locals to manage land between the visits they made from their year-round city homes. This was Queen Victoria's 'tranquil, loyal Highlands' and, although a form of ownership unique in Europe, it continues in Scotland today. Colin's father was skilled in things that ensured esteem in his society: he was builder, shepherd, fisher and crofter, and was thus a fine appointment as deerstalker to a large estate. The family arrived only a day after the bothy's builders left, and were dismayed to find the structure open to the elements. Wind and damp ranged freely through the unlined walls and into the sparse belongings and provisions that would need to last the occupants till spring.

Over the following days the MacDonalds dug blue clay from among the glacial debris at the base of the steep mountainsides, and, with this and planks of wood, they rendered their new home weatherproof.

Gradually, the skies cleared and the grandeur of the neighbourhood was revealed. Looking towards the great mountain of An Teallach (named 'the Forge' for the clouds that hung around its brows long after other tops had cleared) required them to strain their necks: its steep slopes plummeted a thousand metres from a long, narrow summit ridge to their doorstep. Looking south, the valley was spacious and golden with faded grasses. Large rivers wound across it, meeting in wide Loch na Sealga, just discernible over the foliage at its edges. Behind these waterways was a mountain every bit as fearsome as An Teallach, though smaller. This was Beinn Dearg Mòr and beside and beyond it a dozen other peaks fanned out, each unique in its particular brutal grandeur: the horizon in all directions was a stern unbroken barrier of stone. This inaccessibility made self-sufficiency desirable. Vegetables were grown, trout caught, cows milked and sheep sheared. In the hope of reducing reliance on the world across the ridges, Colin's father set to building. Most structures evident beside the Strath na Sealga river today are remnants of his walled garden, drystone barn, sheepfolds and fields.

But some things were always required from outside. Twice a year, people and ponies arrived on the rough track round the base of An Teallach. They carried oatmeal, paraffin, sugar and tea as well as a long roll of cloth from the Ullapool tweed mill. After several days soaked in the river, this was dried on the bothy roof: with luck, an itinerant tailor would pass soon after to cut clothes the family required. There were three other families within the network of nearby glens, and, in order to fulfil the requirements of the Education Act, a pupil-teacher from Dundonnell would spend a month living with each. In the ten years Colin lived at Shenavall, three children were born in the bothy and added to that teenage teacher's charges.

Colin recounted for Alex some tales of bothy winters. Shortly after Christmas 1895, for instance, blizzards brought such snows that it was the end of March before anyone could leave or enter the glen.

One immediate drama was a rescue bid for the sheep stuck on the deep snow of the hillsides. A wide trail was made, like the gap in the Red Sea opened for Moses, through which 500 animals were escorted to the thinnest snow at the edge of Loch na Sealga. One day, when the spring thaw was finally underway, an ear-rending bang filled the glen: the loudest noise anyone present would ever hear. As meltwater ran to the sea, it had left the surface of the loch – a vast lens of ice a few feet thick – suspended high in the air until the weight became too much and a single fissure split the sheet down its six-mile length.

All I could think as I learned these things was that this was an isolation more remarkable than anything I'd heard of on coasts or islands. Although the whims of the ocean could cut people off from others, sea conditions changed more quickly than the turning seasons, and blockades of snow, in regions made remote by clearance, were more impassable than barriers of swell and tide.

Predictably enough, the still, dry weather at Shenavall lasted less than two days before winds grew, clouds gathered, temperatures rose and rain saturated everything. By this time I'd made my river crossings, climbed three Fisherfield peaks and traversed many ridges. Now I was on the lower slopes of towering Slioch ('the Spear') and almost ready to cross the blustery waters of Loch Maree. For the first time for days I was among trees. One reason the Scottish peaks feel like dramatic mountains, rather than the hills they are, is that their oceanic exposure holds the treeline low: elsewhere, the vast, bleak views these uplands offer would be blocked by lush cover of endless trunks and shrubs. Mammals, birds and butterflies, rather than people, would negotiate these peaks and ridges. But even at low level, trees are scarce. The great irony of the old tree cover on this eastern bank of Loch Maree is that this was the spot from which forests for many miles around were felled. In the first phases of the Industrial Revolution, before the processes occurred that clustered mills and factories together and left other regions disconnected, this was a site of prodigious industry. It was a time when Highland land held value similar to that in London (as opposed to the thousandfold differential today) and when the small steps from Loch Maree to the sea permitted integration in the trade networks that counted. Small, moss-covered stoneworks on this wild lochside were once a large and wealthy series of manufactories.

In the seventeenth century, there were three main demands for the wood of this region. Timber merchants took pine and sold it for such purposes as masts for ships. The leather trade began to strip the oakwoods, bark often crossing the sea for use in booming Irish tanneries. And foundries of iron and glass discriminated less by species as they consumed vast quantities of charcoal. Until this time, the Scottish market for glass had been tiny and iron production had been small and localised, conducted on farms or by clan armourers. But on the shores of Loch Maree, a new kind of smelting was pioneered. George Hay was a favourite of James IV who had aided in attempts to colonise the Isle of Lewis. In 1607, hiring English expertise, Hay sank his fortune into a sudden bout of entrepreneurism. He gained royal monopolies in whaling and the glass trade, but his ventures that can still be traced across this landscape were tall blast furnaces for iron. Now just foot-high roots, these sites were powered by the streams that rush from the heights of Slioch and by tons of charcoal from local forests. Nearby ground is full of lumps, betraying heaps of slag and refuse that lie beneath the ferns. With his little 'colony of English' (a term which might apply to their language rather than their place of origin: although one source says they were from Wales and another from Fife, both call them English)

Hay produced iron in unprecedented quantities at three sites on Loch Maree. Though there were rumours of some scandal that chased him from the cities, the forests and mountain burns seem to have been Hay's reasons for choosing this site: his other key resource, iron ore, was mostly shipped from Fife. Perhaps he'd been inspired by the testimony of Timothy Pont who'd passed this way in the 1590s and recorded 'pine, birch, oak, aspen, elm and holly . . . some of the best woods in the West of Scotland'. The prospect of other Scottish ironworks was injured when, in 1609, the Crown banned the use of timber for smelting, but Hay's favour at court exempted him from such petty concerns as the law. It's estimated that, by the 1610s, his furnaces consumed 120 acres of forest a year.

A Victorian historian, John Dixon, collected traditions concerning these ironworks and published them in 1886.[5] Given that, as early as 1950, conservationist impulses towards ancient pinewoods led journalists to describe Hay's actions with phrases like 'reckless improvidence', Dixon's tone of wide-eyed wonder at the industrious energy with which the forests were turned to charcoal is a reminder of how quickly attitudes change:

> How wonderful it seems, that the great iron industry of Scotland, which to this day enriches so many families and employs so many thousands of workmen, should have sprung from this sequestered region . . . the lonely and romantic shores of the queen of Highland lochs.

Indeed, Dixon paints a flattering portrait of a learned, cultured entrepreneur in stark contrast to scornful accounts by Hay's contemporaries who seem to have considered him unprincipled and coarse-mannered. Dixon could recover no knowledge of where the products of the furnaces were sold, though he found many manuscripts that referred to the foundry making 'great guns' and 'casting cannon [until] the fuel was exhausted and the lease expired'. The absence of records means these armaments weren't sold to the Scottish Crown; perhaps they armed Irish fighters or another power overseas.

At least some of the English-speaking workers slowly assimilated into Gaelic life. A small ruin nearby is known as Innis a' Bhàird ('the Place of the Bard'). A century after Hay, a renowned poet lived here: remembered by no name, but as Am Bard Sassunach ('the English Bard'), he was a child of the foundry. The verse he wrote is, like the fate of the Loch Maree ironware, conclusively lost to memory: its only echo is the whisper of place-lore, spoken by the empty landscape.

A few fragments of ancient woodland survived the industrial onslaught: some are thick stands of pine on the many small islands that make this loch so picturesque, but the most extensive are the steep sides of Coille na Glas Leitir ('the Wood of the Grey Slope') on the opposite bank of the loch from the foundries. I inflated my packraft and struggled towards the Coille through winds channelled by mountains into the long groove the loch makes through the uplands. Blown far off the line I'd tried to track, I disembarked beside the first road I'd seen for three days, where interpretive displays advertised the Beinn Eighe nature trails.

Here the remnant of ancient woodland left by Hay inspired key events in the early history of British conservation. After the Second World War, a group of naturalists, all authors for the Collins 'New Naturalist' series, had convinced the British government to found a Nature Conservancy and to seek the foundation of National Nature Reserves for the study and protection of natural habitats, as well as (secondarily) for public engagement with nature. Ancient forests were at the top of their agenda, especially after the privations of war. Coille na Glas Leitir had been greatly reduced in the early 1940s, when Canadian lumberjacks were drafted in to fell two-thirds of the ancient trees. When the war ended, most of the irreplaceable trees they felled still lay rotting nearby. The sudden, almost accidental purchase of the Beinn Eighe estate at knockdown price (around 1 per cent of the cost per hectare of Yarner Wood, Dartmoor, acquired soon after) almost led to all-out war with the Forestry Commission who considered themselves entitled to such a site. But, in a mildly shambolic way, Beinn Eighe became the first official nature reserve in Britain.

Ascending the 'grey slope' it's immediately clear why this small site mattered: the very steepness that saved it from Hay's loggers makes it the most extraordinary habitat. A short climb leads from a mild, damp oceanic climate, sheltered by slopes from the worst of the wind, onto an almost-Arctic quartzite plateau, raked by icy blasts. Each step of ascent brings signs of change. To begin there is lush Scots pine, holly and birch, among which creep pine martens, cross-bills, redstarts and voles. This is the epitome of an Atlantic pinewood, where humidity and extreme rainfall feeds the richest community of liverworts and mosses in Britain: thanks to these veils of diverse organisms every inch of bark looks entirely different from the pines of the drier Cairngorm woods. Once, this was the western limit of a diverse pinewood that stretched round all land masses at this latitude and, in drier, warmer times, reached to the peaks of the hills. Today the pines thin at 300 metres till only small rowans are left. Gradually, woodland birds are replaced by flurries of snow bunting. Patterns appear in the blank soil: the distinctive polygonal networks of frost heave. Soon, even the rowan has gone, replaced by stranger plants lodged in steep ravines. Spiny prostrate junipers are unusual enough, but between them is *Herbertus borealis*, an orange liverwort that grows at only four known sites (the other three in Norway). At the plateau itself, the plants and animals of alpine heights or Arctic tundras predominate: bearberry, ptarmigan, dotterel and mountain hare.

These were the highest hills I'd reached so far on my journey and they revealed most spectacularly the ways in which the immense sweep from ocean depth to mountaintop creates a coastline of staggering diversity and significance, representing a far wider range of habitats than any inland zone on a similar scale. Only after twenty further miles of rough upland, walking over or beneath the long, wild ridges of the great Torridon peaks, did I drop to sea level and cross the water to the tiny village of Shieldaig where, after Christmas with company, I could begin to think about returning to the water.

THE INNER SOUND AND SKYE
(January)

N

KILAMALUAG

Trotternish

Rona

Waternish

Raasay

GLENDALE

Waterste

PORTREE

Inner Sound

S K Y E

The Cuillin

Loch Courisk

Sgurr Alasdair

Sgurr na Stri

ATLANTIC OCEAN

•SABHAL MOR OSTAIG

Kayak route

0   *kilometres*   10

0   *miles*   10

After a glorious, bagpiped Hogmanay in a bar in sight of the sea, the new year began from the northern corner of the Applecross Peninsula. From here I crossed the Inner Sound to Skye and embarked on a long damp journey skirting the island's huge peninsulas. More than anywhere else, Skye is emblematic of the deep histories and scenic grandeur of the north-west. The island's fame is unique: every major Scottish isle has its boat songs, but only one is sung around the world. And Skye has a proud history of defiance against centralising forces that treat all Britain as though it ought to fit a single urbane mould. Its people, along with those of the Western Isles, played central roles in the glamorous tales of Jacobite resistance and led the land agitations of the nineteenth century; its poets, such as Màiri Mhòr nan Òran and Sorley MacLean, have been forces as powerful as armies, able to awake the slumbering force of the Gàidhealtachd in the face of a hostile or disinterested world. Their influence led the first centre of Gaelic further education, Sabhal Mòr Ostaig, to be founded on Skye: when Leòdhasaich or Sutherlanders wish to study the Gaelic past it's to this island they gravitate.

Yet Skye was also the first major tourist centre I reached. It is Scotland's busiest tourist zone after Edinburgh, its every amenity creaking under the weight of seasonal visitors. With Michelin-starred restaurants and multitudes of holiday cottages and hotels, Skye's histories and seascapes are accessible to every kind of traveller. Even at the island's northern corner – one of its most 'remote' and 'unspoilt' spots – I was a few hundred yards beneath a very fine espresso bar: Single Track Café at Kilmaluag. But in the depths of winter, with such wonders shut, it was remarkable how rarely Skye's identity as tourist paradise was evident from the water.

Even the simple act of crossing the Inner Sound meant experiencing Skye anachronistically, because the sea-bound condition of the island has been undone by tourism. In 1995 a bridge to the mainland was funded by private enterprise. With the spending power of tourists in mind it was, per metre, the most expensive crossing in the world. It took a decade for the proud tradition of Skye activism to end the tolls and allow tourists and residents free access to Scotland's most celebrated isle. But the precarious negotiation between visitors and local culture persists: the opportunities and threats of mass tourism are felt more intensely with each successive summer. Every year, new community initiatives such as Eyes on Skye sprout up seeking grass-roots solutions to the burgeoning demands on island infrastructure. When Skye people were canvassed about their perception of current developments in 2016 one fear appeared repeatedly: that the island was becoming a 'theme park'. It's easy to see why that pressure exists. Skye is a Jurassic Park – Scotland's 'dinosaur island' where dustbin-lid-size footprints of lumbering herbivores are nearly three times older than the island's mountains; it's a historical extravaganza, where the romantic tale of Flora MacDonald and Bonnie Prince Charlie grips imaginations; and it's an outdoors wonderland, with climbs and traverses that make everything else in Britain look like a gentle stroll. A handful of tourist sites – 'the big five' – provide visitors with grand castles and fairy glens while barely leaving the road. But Skye reached by sea could – I hoped – conjure different histories from the island by road.

If my Orcadian journey had taken me to the beginning of Britain's built heritage, the sea crossing to Skye threw me far further back. Because of Orkney's slow slide beneath the waves, its most ancient pasts lurk in water. But now I travelled regions where rising land leaves 'relict coastlines' dry, and where traces abound of life before agriculture: Scotland's Mesolithic. The Inner Sound that splits Skye from Applecross is the deepest patch of Britain's inshore waters, but its floor is ridged with banks of rock and sand which form shallows where predators convene. Dolphin pods whisk white water whiter in their frenzied feeding; minke whales (even vast fin whales and humpbacks), porpoises, sharks, sea eagles and hosts of avian rarities compete for what the dolphins leave. On summer journeys I've seen the water round the kayak boil and rip as shoals of mackerel rise and change direction in a single fluid flick. One of the many ridges rises high into two long, narrow islands: majestic, barren Rona divides the northern Sound, while gentle Raasay cleaves the south. Given the right conditions, there is no more magical sea crossing in Britain.

These waters are not just 'whale-roads' but pathways for the most distant of our human ancestors. This is an aspect of their being that was unknown until recently: only in the 1990s were tools found on the Sound's Skye coast that had been quarried by hunter-gatherers who wandered the region 8,000 years ago.[1] Over the subsequent two decades, an extraordinary number of their sites have been located on every shore around the Sound. These finds date from an era when coastlines were remoulded in the aftermath of the last Ice Age. Land rose, rebounding from the weight of ice, but at the same time glacier melt engorged the oceans. Earthquakes accommodated land and sea to their new relations. Not just rock, but forests, herds, shoals and then people moved with the rhythms of the changing earth. Plants arrived in slow, stuttering waves. Trees observed a strict system of social class. First came the 'labourer' trees which did the work that claimed and tamed the land: stunted by hardship and bent double by exposure to the elements, these were dwarf birch and least willow that formed thickets wherever the ice receded. Soon, 'bourgeois' trees set up their shrubby suburbs: hazels flaunted their wealth in rich nuts cast across the forest floor. These trees fed humans like middle-class Britons feed the birds, and vast quantities of burned hazel shell are often found at Mesolithic sites. Last of all came towering tree lords: the oaks and pines. They only deigned to thrust their grasping roots into the deepest earth, and in those favoured spots they flung tall crowns above the canopy.

The remains left by the first human wanderers are small and easy to miss: scattered stone flakes, charcoal, bones of fish, deer, pigs and birds, shells of crabs, cowries, scallops and limpets. Shells were some-times decorated or cut into beads, while many of the mammal bones were worked into points and bevels: tools for gutting, skinning and extracting oils. One such bone, found further south, bears traces of red ochre and raises the tantalising possibility that these people used pigments to make art of bone, body or stone. In this era even peat, now the apparently timeless garb of great tracts of the Highlands, was at the start of its slow growth. It crept down from exposed mountains and seeped up from stagnant pools as it killed the living and preserved the dead. But before the peat's arrival, human artefacts were laid in oxygen-rich soils that worked far faster in providing for life by decomposing death. Early wooden tools are thus rarely found in Scotland, but better-preserved Scandinavian sites yield eel traps woven from willow wands, fish traps engineered from hazel, and canoes gouged from trunks of lime.

Flakes of stone known as microliths were once mounted in wood, bone or antler as tools. Because this stone can often be traced to specific quarries, the social networks of those who wielded it can be surmised. When baked mudstone from Staffin on Skye is found on the Applecross coast it hints at the movements that linked the Sound's sites. The Inner Sound was suited to the needs of people who moved between seasonal resources: low sandy and rocky coastlines, each rich with different shellfish, alternate with cliffs that were filled through spring and summer with seabirds and their eggs. All these terrains are far more sheltered than the Atlantic coasts I'd travelled elsewhere: this has been, for 10,000 years, a friendly place for seafaring. But the land around the Mesolithic Sound was blocked by rocky obstacles and marshland. Dense virgin woodland would have added to the challenges of land travel. Various theories exist for where these humans came from – southern Britain, Ireland, or even the land bridge that once joined Scotland with Scandinavia – but the answer is surely less reliant on land links than was once assumed. The refuse found on the relict shore includes bones of deepwater fish such as cod and ling. These fishers were seafarers, comfortable offshore. Sea travel was not, for them, as unwelcome a risk as it was in the minds of the first scholars to study them. Nor were they simple folk: their arrival on Scottish shores may have been caused by the expansion of large-scale maritime trading networks.[2] As the archaeologist Hugh Carthy put it in 2011: 'we appear seriously to have underestimated the extent and importance of coastal trade routes throughout the prehistory of western Europe, perhaps flippantly dismissing our ancestors as primitive hunter-gatherers'.[3]

This region, with its many shoreline niches for human exploitation, offers a vivid demonstration of the sea's role as the cohesive element of the Mesolithic world. For every onshore structure dug up by Scotland's archaeologists a hundred shipwrecks must lie at the bottom of such places: rowed, sailed and sunk during vast tracts of human history when boats and not buildings were humans' primary tool against starvation and the elements. There are tensions embodied in these scholarly readings of the Mesolithic Inner Sound that are general to the prehistory of 'Atlantic Europe'. Many archaeologists are beginning to see the inshore water of the Atlantic edge from West Africa to Shetland as a 'region' in itself, implying that cultural connections took place beyond the limits of the land. This contrasts the traditional view of the sea as 'barrier': the coast merely the cliff-edge extremity of movements from the inland east. Since a classic text by Barry Cunliffe, *Facing the Ocean* (2001), archaeologists have been more attuned to this issue than writers on later periods where thalassophobic assumption rather than analysis of evidence has often defined approaches.

Yet the Mesolithic is still unrivalled for mystery. Almost all we know about these hunter-gatherers concerns what they ate and how they caught or collected it. We can guess what the world they lived in looked, smelled and sounded like, yet know very little about how they experienced or interpreted their surroundings. They are the only historic inhabitants of Scotland whose world views are almost entirely inaccessible to us: we can barely even speculate on the aquatic gods and monsters, or the pre-Homeric Odysseys, that would enrich our store of stories if these people's mental worlds remained with us. What we do know is that anyone who moves a small boat, the shape of hollowed log, across the Inner Sound is embarking on a journey little changed from the foragings of Scotland's first settlers. As might be expected, it is Norman MacCaig who has expressed this temporal dislocation of sea travel most vividly. The boat need carry no more than a living person, he wrote, 'And there's a meaning, a cargo of centuries':

> Watch this one, ancient Calum. He crabs his boat
> Sideways across the tide, every stroke a groan –
> Ancient Calum no more, but legends afloat.
> No boat ever sailed with a crew of one alone.[4]

There are ways in which all history – everything from ancient Mesopotamia to the present – is the same. Societies underpinned by agriculture share many characteristics, and modern, planned cities such as Houston or Brasilia are, in their most salient characteristics, not all that different from Ur or Damascus. Following the watery threads that lead back to the Mesolithic provokes the realisation that our species is not by nature settled and land-bound. It reveals the myths on which 'we Mesopotamians' have constructed the separation of people and nature from which our current eco-crises emanate: these Mesolithic seafarers have been argued to be the only humans in the whole of time and space who are not the 'anthro' in Anthropocene. And we would benefit from seeing through what we imagine to be their eyes more often.

The long, dark night I spent between knuckles of knock and lochan on the edge of the Inner Sound was intensely atmospheric. I hunkered down against a thin smurr of rain, sometimes caught in moonlight, with the thick smell of sodden peat eclipsing the salt of sea just feet away. And I read about the most celebrated boats to have plied this water. The book I read, Alasdair mac Mhaighstir Alasdair's *The Birlinn of Clanranald*, is one of the great Gaelic seafaring epics: an Iliad in which the Troy to be stormed is this Hebridean sea itself. Written in the 1750s, it's set at a timeless point before Culloden, when islanders still wore kilts and chain shirts: its symbols often seem to belong to the fifteenth and eighteenth centuries simultaneously. The author was a Jacobite who commanded fifty men and tutored Prince Charlie. When Hanover triumphed at Culloden he left the mainland for the Hebrides to escape recrimination for the scathing verse he'd aimed at the new royals. His world remained that of the seafaring clans MacDonald and Clanranald: the north of Ireland, Argyll, Islay, Uist, Canna and Skye.

The birlinn was bigger than the sixareens of Shetland, comprising twelve to eighteen oars and a square sail. Although clinker-built in the Norse tradition, it was a further step removed from Norway, not double-ended but with a flat stern to permit a steering oar or rudder. Sailing seas north from Ireland, birlinns became a currency of liege and lordship: the number of galleys a clan could muster defined its prestige. The birlinn is therefore immortalised on clan crests and the walls of coastal chapels such as Rodel (Harris) and Rob Donn's Balnakeil. Just as the culture of Sutton Hoo dragged boats up hills for symbolic burials, the societies of these islands brought the sea ashore, placing symbolic ships at the centre of their towns, castles and churches. In this way, the birlinn became an icon of the Atlantic ties that bound Ireland, Man, Argyll and the Hebrides. It recalls cultural formations, such as the Lordship of the Isles, that show Scotland – like England, Wales, Ireland and Britain – to be an idea moving through these islands only a little slower than a ship at sea. Before these nations, each only really united by modern legal codes, there were, for millennia, loose confederations of multilingual, multi-ethnic interest groups.

Tradition holds that, seeking inspiration for *The Birlinn of Clanranald* while he was baillie of the isle of Canna, Alasdair lay beneath an upturned vessel on a Hebridean shore. Entombing himself in darkness, with only the smell of boat for company, was a strategy to spark imagination. This principle became an *idée fixe* among Atlantic aficionados. Ludwig Wittgenstein, for instance, channelled Alasdair when he claimed to 'only think clearly in the dark' and, in 1948, fled the street lamps of south-east England for waters the birlinns had travelled: he noted, with approval, that the Irish Atlantic he found was 'one of the last pools of darkness in Europe'. Seamus Heaney, at his most elemental and earthy, wrote himself into this proud tradition. The final lines of 'North' are set on a long strand with only the 'secular powers of the Atlantic thundering'. The sea inspires reverie that sends the poet spiralling back centuries to see the water as the road of Norsemen. The 'swimming tongue' of a historic longship speaks to Heaney and invokes the poetic darkside:

Lie down
in the word-hoard, burrow
the coil and gleam
of your furrowed brain.

Compose in darkness.
Expect aurora borealis
in the long foray
but no cascade of light.

Keep your eye clear
as the bleb of the icicle,
trust the feel of what nubbed treasure
your hands have known.

It is perhaps surprising that the poetic fiction born of Alasdair's self-imposed enclosure contains such detailed description of the birlinn's structure and the actions of its crew. It is the best evidence we have for the facts of what this vessel was. No examples of the boat – even wrecked – survive: in 1493, when James IV absorbed the Lordship of the Isles under the Scottish Crown he demanded that all birlinns be burned to end the power of the sea lords.

Alasdair's birlinn moves through a Hebridean sea that's as cunning and wise as human or animal. It's an old man with streaming grey hair and a creature with gaping jaws and matted pelt.[5] As a respected foe, the sea's will is pitched against the desires of the boatmen. It responds to being struck with oars until, eventually, it submits to human strength. The boat is also alive, crying out like a person and whinnying like a mare, treading waves not with planks and thwarts but shoulders and thighs. Boat and boatmen are one: the sweat on the sailors' brows is the brine foaming round the bow. And the boat becomes their homeland as they climb creaking mast and ropes 'as quickly as May squirrels on the trees of a dense forest'.[6] At sea all distinctions between animate and inanimate, sentient and insensible, human and animal, flounder. In these verses, as in much writing on its waters, the Minch is layered with metaphor; the inter-island seas are known like friends and rivals; waves and tides are feared or loved like animals of hill and forest. Here is humanity engaged in the quest for mastery over nature: for separation from the seething conflicts of the bestial, elemental world. But to Alasdair's protagonists, before the age of steam and steel, that quest still seemed impossible; dividing lines, distinctions and disentanglements can rarely survive a single line of verse.

Next morning, I prepared my own encounter with the grey-haired sea in mist that made me alert to animal encounters. Before I even hit the water, a brute of a dog otter surfaced on its back, scarred snout and crab catch raised above the waves. It didn't bother to acknowledge my presence but rolled like a thing uncoiling, then lolloped noiselessly into brown remains of bracken. It took seconds from its departure for its passing to feel mythic, and moments later I was moving through cold smoke-like rain towards a lunchtime landing beneath the Rona lighthouse.

This night in the fog had established the tone for the month. As I crossed the Inner Sound and kayaked each long finger of Skye's western edge I breathed mist, drifted through sweeping rain, and saw the island only as shape-shifting cliffs that loomed, suddenly, from saturated skies. Headlands were bands of thick dark haze, and I found I could judge my distance from them not by their size but by the degree to which they blackened the otherwise featureless pall of grey.

The otter felt like an appropriate sigil of this place because it has long been treated as hybrid and unknowable. Like the barnacle goose, otters were a conundrum for the monkish administration of Lent: both seemed more fish than bird or mammal. Some Carthusian monks were forbidden meat all year round. Instead, they ate otter. In Norse and Celtic story otters, particularly otter kings, change form and grant wishes, but only in the unlikely event of their capture: the animal's fluidity gives it the character in water of intangible smoke in air. The otter is its element: 'ninety per cent water', to the poet Kenneth Steven, and 'ten per cent god'.[7] But they are also friendly 'water dogs'. They brought St Brendan fish and firewood; they warmed and dried the feet of St Cuthbert when he finished his nightly vigils waist-deep in sea. In the work of the great poet-naturalist Colin Simm the otter is a boat that's 'all rudder'; it is Mesolithic, belonging in ice melt 'a few thousand years back' when elver-silvered rivers still thronged the landscape. Simm has written hundreds of closely observed otter poems, and in many, floods are the creature's medium. Water sweeps land when, in acts of drainage and deforestation, 'a balance of centuries to the balance-sheet yields'.[8] When otters twist and tumble through redrowned vales a historic ordering of water, earth and animal is reprised in a beautiful unplanned catastrophe of rewilding.

As poets make otters into ribbons of water, so they make Skye a figment of fog, a realm subject not to divine or human law but to 'amorphous rules of light'.[9] When Richard Hugo, poet of the Pacific Northwest, came to live on Skye he wrote that the shifting mists alter the colour of the island a hundred times a day and 'never stop changing the distance to the pier from your front door'. Skye's epithets – to the Norse, Island of Cloud; Misty Isle to the Gaels – are aerial and never earthy. The prevalent sou'westerlies are 'the grey wind' that scoops the otherworld of the sea ashore. This island is the grand centrepiece of the Hebridean world, straddling the Minch both north–south and east–west. Smaller than the land mass of Lewis and Harris, its coastline is far longer: its gangly peninsulas intercept fog-bound vessels on a hundred different inter-island routes.

Skye's geography has long been mystified: it is '60 miles long', according to the mountaineer W. H. Murray, 'but what might be its breadth is beyond the ingenuity of man to state'.[10] This is perhaps why Skye is the most zoomorphic of landscapes: an animal island. When factual delineation falters on its ragged edges, allusions to diverse living things scuttle in. Some are merely evocative: 'Skye sticks out of the west coast of northern Scotland like a lobster's claw ready to snap at the fishbone of Lewis and Harris.'[11] But other Skye animals are potent political symbols, as 'the gull' and 'the stallion' became in the hands of Skye's greatest cultural icon, Sorley MacLean.

MacLean was a colossal presence in the twentieth-century Gaelic cause and both these animal images appear in the ferocious 1930s poem that voiced his rallying cry for a new island politics, *An Cuilithionn*. For MacLean, the things that define the island – land, sea, society and animal – were interlocked and each is illuminated through imagery of the others. The jagged peaks of the Cuillin are, for instance, 'a great dim sea of gabbro waves' and a mobile world of rising, 'bubbling crags'. In other poems, human sickness and death are figured as the stilling of tidal movements: 'dead stream of neap in your tortured body / which will not flow at new moon or full'.

In the figure of the gull, MacLean links Skye's past to its geography. That Skye is a bird is suggested by its wing-like peninsulas and the probable derivation of its name from '*skitis*', a pre-Norse Celtic word for 'winged'. Yet the seabird Skye in MacLean's usage is not an emblem of soaring freedom: it has been wrecked by the storms of a cruel history and is floating prostrate in the Minch: 'Pity the eye that sees on the ocean / the great dead bird of Scotland'.[12] The gull in the storm evokes the impossible odds against which islanders have fought in a political system structured to mainland needs and the profit of landlords rather than land-dwellers.

MacLean's other image, the wild stallion, links Skye geography to the vigour and resilience with which islanders have resisted the impositions of distant power. It is an icon that might, at first, appear land-bound. But the stallion, MacLean scrawled in his notebooks, was inspired by a cliff known locally as Stallion Rock: it is 'the magnificent sea cliff at Waterstein used as a symbol of the heroic conception of Skye in Scotland'.[13] As the horse runs it is 'craggy' and 'rocky', always retaining its identity as a feature of Skye's coast. The stallion evokes particular moments in island history. In 1883, the people of Glendale, a coastal township not far from the stallion cliff, took action against the landlords who curtailed their land rights. In 1883, rushing down from hillsides, cudgel-wielding crofters clashed with police at a bridge on the Hamra river. They demanded restoration of common grazings and coastal customs, such as the right to collect reeds for thatch and driftwood for burning and building. After the police were rebuffed, distant urban authorities lumbered in: a naval gunboat was despatched to cow the handful of maltreated crofters around the bay into submission. Five Glendale men stood trial in Edinburgh and were soon seen as noble martyrs in contrast to a brutish officialdom. Thanks to fierce and eloquent Skye poets such as Màiri Mhòr nan Òran ('Big Mary of the Songs') the cause of the 'Glendale martyrs' and another 1880s moment of unrest, 'the Battle of the Braes', was quickly sung across the global Gàidhealtachd.[14] These insurrections proved a turning point in the crofting struggle, when public opinion and the views of the prime minister, William Gladstone (a Liverpool-raised politician who always liked to believe himself Scottish), finally moved in the crofters' favour.

MacLean's stallion is the hope engendered in and by the movement Big Mary had sung half a century earlier. In the stallion's galloping, the very rocks of Skye's shore rise in sympathy with islanders. The stallion is not a generic symbol of hope, but a specific anti-capitalist and anti-fascist energy. MacLean saw island life as intrinsically socialistic. Gaelic is *Gemeinschaft* (organic community) as opposed to the *Gesellschaft* (commercial society) of English; it is coexistence between humans and their world as opposed to exploitation of the poor and the environment. As the stallion climbs the peaks of the Cuillin, the ascent to airy heights from its world engulfed in waves signals the freedom engendered by a global outlook. The lone stallion sees other 'mettlesome strong horses' resisting 1930s fascism across the seas in Russia, India, France and China. By the time MacLean penned an introduction to his collected essays, Stalin had disabused him of his youthful enthusiasm for communism, but he elaborated the links he'd perceived between local and global events: 'in 1938 the continuing existence of Gaelic as a spoken language seemed a forlorn hope and Europe itself appeared about to be delivered into the hands of Teutonic racist fascism'.[15] MacLean's Gaels were, by their very nature as serial resisters, Britain's radical front line against cultural chauvinism in Europe's darkest hour.

When I kayaked beneath Stallion Rock, I was looking up at a remote inaccessible sea cliff physically unshaped by human action. That this blank rock became so powerful a part of Skye's cultural landscape – an ancient natural monument to a modern political moment – is profound illustration of the diverse ways in which humans leave their marks upon coastline.

Had the cloud lifted, the Cuillin hills would have become more domineering with every mile I travelled south down Skye. As it was, the highest spectacles were ribbons of cascading water tumbling from the black clifftops. Great northern divers, known as storm fowl, cackled day and night. I explored the small-scale structures of the coast, finding evidence that Skye's long limbs straddled not just the Minch but the world. Graves in coastal churchyards pointed to India and North Carolina, while one son's stone memorial asserted his imperial status as he commemorated his mother:

Sacred to the memory of AGNES PINKERTON SHAW daughter of the late Reverend JOHN SHAW of Bracadale and widow of the late Reverend JOHN BROWN of Cray Glenshee who died 23rd November 1884 aged 62 years. Blessed are the dead who die in the Lord. Erected in loving memory by her son JOHN SHAW BROWN Municipal Engineer Akyab Burmah.

At least five men named John Shaw Brown linked Akyab with the coasts of Scotland and Ireland: the name crossed several branches of one ambitious imperial clan. Boats bearing linen made by a Belfast John Shaw Brown didn't pass Skye's Stallion Rock or any foggy Scottish shore, but instead witnessed crocodiles and tigers from paddleboats that shipped Burmese teak along the sunlit Irrawaddy river.

By the time I landed in darkness and surf where the Cuillin meets the sea, it had been four days since I'd been dry and far more since I'd seen the sun. Next morning, cloud still veiled everything and I resigned myself to another damp shoreline day. But suddenly, a patch of cloud thinned and whitened. Through it flickered a clear black peak. Tripping over my feet in eagerness, I began a fog-bound trot to the high corrie of Coire Lagan, where a serene loch nestles in an amphitheatre of splintered gabbro and other glacial ruins. Where the loch forms a burn and begins its long seawards tumble there are traces of a small stone wall: this corrie was the highest and most remote of summer shielings and the stones were placed by its shepherd to keep his goats from tumbling from their home on this surreal ledge. The shieling belongs to an era before the supposed first ascents of the surrounding peaks, yet it seems unlikely that the unknown shepherd never pursued a roving goat among the jagged spires 1,000 feet above.

However improbable it might at first appear, there are many ways up the mountainous walls of Coire Lagan: I waded up a precipitous wall of loose volcanic scree known as the Great Stone Shoot, sliding five steps back for every ten upwards and releasing musty ancient fragrances from the stones. The Cuillin can be a strangely seasonless world, bitterly cold in August and gently warm in winter, its character defined by the shifting ocean winds. The ptarmigan and plovers whose changing appearance marks the turning of the upland year are absent here, as is the sward of moorland plants that elsewhere greens the bright months and browns the dark. The jagged, oceanic peaks hold snow for fewer days than other Scottish mountains of their height. Today the sun was dazzling and the air still: though it was January below the clouds, it felt like June on the snowless slopes above. I reached the highest peak, Sgùrr Alasdair, by mid-morning and looked out not on the tapestry of coasts and islands I'd anticipated but on a perfect inversion: a total canopy of low white cloud (figure 7.1). From An Teallach in the north to Ben Nevis in the south, the mainland mountains rose from billowy skirts. Some hills, such as this range's smaller sibling, the Cuillin on the Isle of Rum, didn't pierce the clinging vapours. Instead cloud massed on them until they looked like icy burial mounds in a prehistoric snowscape. Around the ridges I'd walk, cloud flowed like water over beallachs that blocked its slow northward course. This was a vision of what Britain would be with sea levels raised half a mile with mountain peaks as isolated islets.

There's barely room to sit on Sgùrr Alasdair's summit, but one of the first people to scale the peak, the renowned Alpine mountaineer Charles Pilkington, solemnly requested of each climber who followed that they give the fine views 'at least an hour of [a] misspent life'. Pilkington warned, however, that the inexperienced climber should not be here alone: 'Sgùrr Alasdair, though not so high, is a true mountain, and not a hill like Ben Lomond or Skiddaw.' This is, indeed, the only mountain range in Britain to make the seasoned alpinist's pulse run too quickly for comfort.

I was joined before long by a Yorkshire mountaineer who, like many climbers, had a total loyalty to the Cuillin: bypassing the rest of Scotland, he comes from Halifax each year to climb with Skye's mountain community. We stayed on the peak, as Pilkington requested, talked mountains and watched the cloud level sink, until we each went separate ways. He returned for music at one of Britain's celebrated mountaineering haunts, the Old Inn at Carbost, while I wandered the climbers' playgrounds of the ridges, the cloud turning pink and gold as a huge sun melted through it. By evening the cloud fell so far that low landforms I'd passed on my first day of paddling – the Stoer and the Quairang – stood out above it in the distant north. This was one of the most beautiful nights of my life, all the more powerful for the drab days that preceded it.

By morning, January was restored; the wind had risen and temperature dropped dramatically. Cloud swirled above the tops, only gradually forming a blanket at eight hundred metres. With fingers scraped raw by gabbro scrambles and biting wind, I looked back to Sgùrr Alasdair from Sgùrr na Banachdaich. Peaks with names like Sgùrr Thearlaich and Sgùrr Mhic Choinnich appeared at intervals from speeding billows between. The history of this skyline vocabulary is unique in Scotland, because so many names were altered in the half-century after 1870. Sgùrr Alasdair was called Sgùrr Biorach ('the Pointed Peak') until the day in 1873 when the Skye poet Alasdair Nicolson was first to conquer a summit previously thought inaccessible. Nicolson was sheriff in Skye's main town, Portree, and used his sheriff's plaid as rope on tough manoeuvres. Sgùrr Mhic Choinnich immortalises a mountaineer, John Mackenzie of Sconser, who was climbing 'virgin' peaks by age fourteen; Sgùrr Thearlaich ('Charles's Peak') is named for Charles Pilkington, the Merseyside climber after whom Mount Pilkington, British Columbia, is also named. No other range in Britain lost its local identity to recreational names in quite this way.

The story of the opening of the Cuillin by these pioneers is the story of the opening of Skye to visitors of all kinds. It began with the birth of a new aesthetic, climbers following in the footsteps of poets and painters. Early tourists didn't approach these mountains from the land as modern visitors do but from the water, entering the Cuillin bowl from the salt loch to the south and disembarking to explore the shores of freshwater Loch Coruisk (Coire Uisg – 'the cauldron of waters'). By this route well-to-do tourists made Loch Coruisk Britain's national emblem of romantic sublimity.

Walter Scott was among those who extolled the scene. His 1814 journal recounts a journey by rowing boat to the bottom of the 'huddling and riotous brook' which drains Loch Coruisk into the sea, where hundreds of trout and salmon were struggling upwards. He described the 'exquisite savage scene' where huge strata of naked rock 'as bare as the pavements of Cheapside' rise 'so perpendicularly from the water-edge, that Borrowdale or even Glencoe, is a jest to them'. The effect was desolation more potent than Scott had ever experienced, but 'its grandeur elevated and redeemed it from the wild and dreary character of utter barrenness'. Scott featured the Cuillin in his poem *The Lord of the Isles* and commissioned no lesser an artist than J. M. W. Turner to illustrate the scene. Turner too visited the loch although unlike Scott he braved a climb above it, mounting the shoulder of Sgùrr na Stri, a small but spectacular hill on the seaward side of the Cuillin range. Here, he slipped and only 'one or two tufts' of Cuillin grass prevented a fatal fall. Turner's sketches focus as much on watery depth as rocky height, illustrating Scott's description of Loch Coruisk as a 'sable ravine and dark abyss'.

Scott was obsessed with the Jacobite legacy, and Prince Charlie's flight across Skye became another means by which early travellers could imagine the island as the last stronghold of authentic Highland spirit. Scott's heady fusion of history and landscape formed the basis of tourist expectations, and those who followed in his footsteps soon multiplied from dozens to thousands.

The new media that draw many modern visitors to Skye derive directly from the aesthetic built by Scott and Turner. The historical romance of television shows such as *Outlander*, which opens with the Skye Boat Song, is descended from Scott's historic novels. Car adverts and music videos (such as Harry Styles' 'Sign of the Times') feature sweeping views of Skye's landscape inspired by Turner's sublime. Our own, twenty-first-century moment seems oddly captivated by the aesthetics and interests of an era exactly two centuries earlier. But a wholly modern ease of travel and communication means that Skye is no longer insulated from the implications of its deserved celebrity. Every ground on which that celebrity rests – from the well-preserved historic sites to the wildness of the land in which they're set – is threatened by precisely those who love them most. And the island, in its need for tourist income, is sustained by the dynamics of its own destruction. The condition – insoluble but unignorable – is Shakespearian in its many forms of grandeur, pathos and tragedy.

ARGYLL AND ULSTER
(February/March)

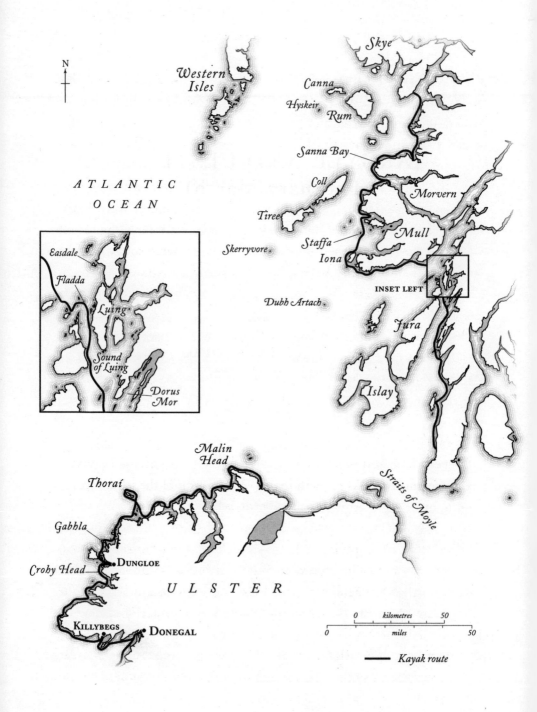

N

*Western
Isles*

Skye

Canna

Hyskeir

Rum

Sanna Bay

Morvern

A T L A N T I C

O C E A N

Coll

Tiree

Mull

Easdale

Fladda

Skerryvore

Staffa

Iona

Luing

INSET LEFT

*Sound
of Luing*

Dubh Artach

Jura

*Dorus
Mor*

Islay

*Malin
Head*

Thorai

Straits of Moyle

Gabhla

Croby Head

Dungloe

U L S T E R

0   kilometres   50

KILLYBEGS   **DONEGAL**

0   miles   50

—— *Kayak route*

O NLY IN FEBRUARY did winter truly hit. Finally, the hills were shrouded in snow. Tides like tumbling alpine rivers rushed me southwards along narrow channels of an inter-island world. It was ten days until, beneath the southern cliffs of Mull, I lost sight of the snow-topped Cuillin: for all that time Skye's mountains stood proud against blue-white skies while brisk easterly winds held dark fug offshore. Stony beaches where I landed were laden with thick, rubbery stalks of brown kelp: storm salads recalling recent gales. But the sea twitched with the memory of fierce sou'westerlies long after the wind turned east: beneath Rum, and along the rough Ardnamurchan coast, high breakers suddenly appeared even when the sea looked safe.

These stretches of coast have dark histories. They're places where even visitors from Skye have felt flayed by loss. In 1937 Sorley MacLean moved to Mull where he expected comfort and a sense of homecoming at the hearths of clan Maclean. Instead of calm he felt rage:

> I believe Mull had much to do with my poetry. Its physical beauty, so different from Skye's, with the terrible imprint of the Clearances on it, made it almost intolerable for a Gael.[1]

Mull's history of emptying is long – much of it pre-clearance – and devastating: the Gaelic community here is but a fragment of those further north. Most Mull residents I met were English or Irish. Some were ocean scientists (most nations have fewer cetacean specialists than the Isle of Mull); others were outdoors guides, or wildlife photographers. The island's fauna has become its defining feature in large part because the main roads skirt sea lochs: nowhere else in

Britain can otters and eagles be so reliably observed from the comfort of a car. I also met kayakers here in numbers greater than elsewhere. In many ways, Mull is already the theme park that Skye folk hope their island won't become.

Mull's neighbouring mainland peninsulas are sites of two of the great accounts of cultural loss. One of these, Philip Gaskell's *Morvern Transformed* (1968), is among the most significant histories of clearance: an early study that broke with historians' tradition of open sympathy with the landlords. The other, Alasdair Maclean's *Night Falls on Ardnamurchan* (1984), is a personal narrative of communal traditions slowly attenuated. Maclean's book is a dialogue with the diaries of his father who had been a crofter on Sanna Bay. The father personifies measured reserve through his bare descriptions of change; this is paired with the son's emotive elaboration of the consequences of that transformation. The two combine into an understated account of the loss of ways of life that carries exceptional force. Such texts cannot but colour a kayak through this region in winter, when holiday homes are empty and the distance to any shops and facilities immense. I thought back to my journeys down the great archipelagos from Shetland to the Hebrides. The islands had rarely, if ever, felt truly remote: there were always well-provisioned communities relatively close. On mainland coasts such as Sutherland and Ardnamurchan, isolation and lack of investment feels more persistent, and the task of staying well stocked in the kayak loomed larger.

Winter departed as swiftly as it arrived. March brought fine, warm weather, the sudden eruption of flowers in cliff-face fissures and even the exceptionally early arrival of puffins. By the time this turn took place I was among islands that felt similar, yet exist in a different political order. The twenty-six counties of independent Ireland were only ever coerced into the political entity of Britain during an extended nineteenth century (1801–1919). This journey across national borders brought the best opportunity to consider the integrations and divisions of this archipelagic world.

It was neither the large island of Mull, nor long headlands like Ardnamurchan, that defined my perception of the bleak February weeks at sea. The list of small isles passed – from Canna to Cara – could fill a page. The smaller the isle the more it seemed to shape my sense of the history on my journey. Many were long depopulated, some were still home to communities in triple figures, a few to populations that have long been confined to tiny numbers only. The experience of these islands reprised and reinforced a host of the themes I'd witnessed in the Northern and Western Isles. As on islands further north every community here required extraordinary resilience to survive the years of rampant modernisation after 1780, triggered by the events of the so-called 'Enlightenment'.

I thought as I travelled about how that period demands rereading in the light of island stories. I thought of some of the wisest words ever spoken about history, said by Calvin, the child protagonist of the cartoon strip *Calvin and Hobbes*: 'History is the fiction we create to persuade ourselves that events are knowable and that life has order and direction.' In the 1970s concepts similar to the Enlightenment – labels that implied a direction in history – were revealed as little more than propaganda praising certain interest groups. Feminist historians such as Joan Kelly-Gadol, for instance, showed the 'Renaissance' to be a narrative that fits the experience of a cadre of wealthy upwardly mobile men, but not their contemporaries whose opportunities narrowed and wealth decreased.[2] To sum up an era with the term Renaissance is thus to engage in an identity politics that values the rich alone.

The case of the Enlightenment is little different. Social distinctions of race, class, gender and sexuality were not undermined but consolidated: this was the era of scientific racism or 'the century of the colour line' as it was labelled by the philosopher W. E. B. Du Bois. Yet the case against the label 'Enlightenment' is also a geographic one: to deploy it is to be dazzled by cities and blind to rural sea coasts. As scholarship informed by environmental challenges increasingly encourages focus on place and geographical distinctiveness, the Enlightenment must surely fail as an explanatory narrative.

Every island's story of negotiation with the centralising forces that would have stripped it of wealth, culture and people, was unique. A kayaker who lands on Easdale wanders into a realm that's both post-industrial dystopia and unspoilt wilderness, every thistle a reminder of how rapidly and unpredictably nature's domains are wrested free from human domination. Slate waste pits make deep blue pools near a bright modern community that resides in clean, whitewashed ex-miners' cottages. Landing on the west coast of Rum is an entirely different prospect. Nestling beneath bleak mountains are a striking array of abandoned lazy-beds and a crassly urbane Graeco-Roman mausoleum. The similar forms of prostrate agricultural ridges and upright architectural pillars are the same histories refracted through poverty and wealth: a whole way of life is buried unceremoniously in the ridged earth, while one wealthy owner is immortalised with bizarre ostentation. Rum was bought and sold across the era of Enlightenment and its aftermath by families whose wealth came from Caribbean plantations and whose kinship groups half-included children born from the rape of slaves. Given the emotional calluses built over lifetimes of imperial exploitation it's no surprise these landowners failed to see worth in ways of life they obliterated in island glens.

Such suffering was now familiar. But here, unlike on other coasts, I was treading the terrain of one of the great political campaigns that drew general attention to the plight of the islands. In the early twentieth century, the projects of state building that had defined previous decades continued, metropolitan visions of the nation gradually swamping regional or local patterns of organisation. But resistance also began to gather momentum. On coastlines around the Atlantic protests took the form of long sea journeys that were used by coastal communities to corral new media into publicising the plight of places and trades that the nation seemed to sideline. In Brazil, for instance, several voyages were made, between 1920 and 1950, protesting government failures to recognise fishing as an official trade.[3] The only such journey to take place off British coasts was also the one such voyage to be conducted by kayak.

In 1934, Alastair Dunnett and Seumas Adam were young journalists, raised in the tenements of Glasgow, who'd lost every penny of their meagre savings in their creation of a boys' adventure periodical, *The Claymore*, which ran for a year before its debts caught up with it. Dunnett and Adam had met through outdoors pursuits at the moment when, as Dunnett's son, Ninian, put it,

> The hordes who had flocked to industrial Glasgow to bend their shoulders for 'the second city of the Empire' were beginning to rub their eyes and discover the fabulous landscape on their doorstep.[4]

Unlike other industrial or clerical workers, who explored the Highlands by land alone, these two took frequently to the sea, heading to harbours at dawn and hitching rides on coastal cargo tugs or herring drifters bound for the Hebrides.

Those improvised journeys inspired the spirit of *The Claymore*. The burgeoning market of boys' adventures, Alastair Dunnett noted, was almost entirely imperial in its content, featuring 'South Sea planters sons' and 'pet gorillas which played cricket'. If working-class boys appeared in these stories they were 'bad yins': 'coarse errand lads' whose many 'villainies were mere by-products of the major crime of working for a living'.⁵ *The Claymore* aimed to show Scots that children could have adventures without needing to attend an English public school or be posted to the colonies.

When *The Claymore*'s publisher at last grew tired of their accumulating debts, Dunnett and Adam sought a new venture through which to pursue their cause at little cost. Through *The Claymore*'s advertisement pages they had built relationships with outdoors industries of every kind and had discovered in Queensferry a man named John Marshall who'd begun to build canoes from canvas, rubber and teak which he insisted 'would take a skilly man anywhere in the waters around our coasts in all but the most severe conditions'. These vessels gave the duo the idea of kayaking from Glasgow to the Western Isles as a voyage to explore their heritage (Dunnett's mother was a Gaelic-speaker from Loch Fyne, and both men traced their heritage to the Highlands and Islands) and the latent potential of struggling islands economies. On the day they set off, the papers published their press release aimed at all the people of Scotland:

> Our intensive industrialisation has lost us our contact with the sea. Steamers have made us forget the thrills of small-craft sailing. We want to test the zest of physical living that town life denies us. But there is something more . . . We want to see the Western Isles thoroughly, to complete our own picture of them. Not the Isles of the guidebooks, but the real isles – the isles of Opportunity, peopled by a vigorous race with an unrivalled climate for some types of products; islands capable of supporting more of our surplus population in large-sized holdings yielding an adequate return. Therein is our real adventure – exploring the possibilities for expansion and development in our own country.

Most importantly, as Dunnett put it later, 'it was the land and the people, and not we, which would make up our story. They would take the foreground, and we would try to make ourselves a part of that scene and life.'

The Queensferry canoes (what Dunnett and Adam were up to would be referred to as kayaking today, but the terms canoe and kayak were used interchangeably until later) were rudimentary vessels, even involving an inflated car-tyre inner tube as a cockpit. Their other equipment was no more specialised, consisting of kilts, tweeds and long thick woollen socks. Their preparations were as improvised as their vessels. Although they claimed in their press releases that 'both are, of course, expert canoeists' neither had in reality 'even seen canoes at close quarters until a few weeks previously, and our only experience of handling them had been on a recent Sunday afternoon on the Forth and Clyde Canal'. Most of their ocean-going knowhow came from a chance encounter, in the gym, with a master mariner.

Setting off in August 1934, they reached Skye before turning back, having encountered oceanic challenges of every kind, and provided gripping descriptions of battles with tides and swell. In one set piece, for instance, their mariner adviser had warned them to tackle the tidal phenomenon off Craignish Point known as the Dorus Mor ('the Great Door') at slack tide. But after being told by some overconfident herring fishermen that taking it during the flood tide would speed them north to Oban, they found themselves dragged into a waking nightmare:

The water changed in colour from a pleasant green to a sudden and sullen black, in which writhed streamers and trails of spent foam. And with the colour went the one-way rhythm of the water which had taken us here. The lifting waves that had followed and passed us in reliable attendance were drowned in a jauping popple.

These separate wave peaks reared individually and fell on us, punching at our sides and canoe tops, each one jerking us as solidly as a thrown bucket of stones . . . Dipping paddles, and tugging and staring at the near water, we appeared to be fixed and struggling in a static maelstrom. A glance farther ahead corrected the impression, for onwards and approaching, was a low wall of water, higher than the level we were on, where the two irregular tide forces were heaving up the sea between them.

The wall seemed to dart and strike us, although it was we who rushed on it. Here the paddles felt new forces that made them kick in our grip as if hands in the water had seized them to wrestle them from us. We were now in a moving group of whirlpools, and the noise was a hissing thunder. On the other side of our hulls of cloth and slats the sea gathered below our thighs like a horse bunching for the gallop. I struck the perimeter of a great swirl, swooped half round it and rammed Seumas with my point on his bow, remorselessly, although we were both stroking fiercely apart. We clashed together for a moment along the length of the hulls, and parted on our ways again. Several times there would come a sudden subsidence of the water, leaving one or other of the canoes sliding on the surface of a smooth bubble platform of sea, 20 yards across, pressed inches higher than the surrounding level like a lily-leaf adrift. Then this would burst and rip and the spouts would storm at us, and a force below would seem to twitch the canoes deeply down below what buoyancy still ruled them.

After describing the whole horrifying sequence, Dunnett writes simply 'there is an elation which follows novel perils'.

Besides their balance of humour and lively writing, the canoe boys' greatest skill was their handling of the press. That they knew the industry empathetically and intimately is indicated by the fact that Dunnett soon occupied its highest offices, campaigning for coastal development and devolution through twenty years as editor of the *Scotsman* before he was knighted for his public service. Sending frequent missives back to the *Daily Record* and the *Evening News*, they balanced the derring-do of articles such as 'A Battle With a Gale' with the serious social description that, after the journey, was compiled into a seminal longer article, 'A New View of the West', which set out their dreams of revived local industries that undoing a century of centralisation might achieve.

They hadn't gone far before the publication of the first article gave them crowds of curious well-wishers at every Hebridean port; and pausing in towns such as Tobermory they were able to take the temperature of local feeling. There were some who, having tasted too much 'of the sponge dipped in vinegar which is constantly presented to the Highland mouth', offered only pessimism and discouragement; but most were 'kindly and keen' with a belief in the Highland spirit to reverse the present hard times.

Dunnett's writing paints pictures of diminished but vigorous communities in places like Mull. This was the home of the Macleans: the cradle of the clan. Yet there were only ten Macleans in the local phonebook. There were more Macleans than that in the Buenos Aires phonebook (eleven), never mind the 155 in Manhattan (an island he describes as 'about the size of Colonsay'), the 320 in Glasgow and the 370 in Toronto. At Ardmore, opposite Mull, the canoe boys found themselves among ruined crofts. 'If I am told,' wrote Dunnett, 'as one tends to hear in such cases, that these people were able to scrape only a wretched living from the ground, and were better dispersed on their emigrant business, I shall require proof more persuasive than is usually available.' He lists at length the fruits of land and sea that must have 'given a hearty fullness' to the old Ardmore days. But the many crofts had been replaced with a single Victorian farm in order to increase commercial profits from the land and, as Dunnett put it, 'the solitary house, and the system of husbandry and lairdship it signified, had not even a hint of the endurance of the community that had been there; and that might, if management and understanding had been nearer home, have been there yet'.

This destruction and abandonment and exile was a theme the canoe boys saw everywhere: they travelled at the end of a century-long era of heavy depopulation. But it wasn't the whole of their story, because it didn't account for the providence of the land and the vigour of those who were left. The thriving shops of Tobermory impressed them: Coll cheese, Barra potatoes, carrots from Muck, ling sun dried on the rocks, as well as Mull honey and gooseberries. Their greatest pleasures, however, were always the most practical: 'what delighted us most about the place at that time was its recently completed and self-contained hydroelectric scheme'. If a small town could create its own electricity supply, jumping overnight from paraffin lamps to electricity, then everywhere in the islands must be capable of forging its own forward path.

And their highest praise of all came for the independent people outside towns, such as the single young family who farmed Calve, the island in Tobermory Bay. Here Dunnett and Adam stopped to assist with the harvest. An old traditional storyteller named Locheil made frequent visits to their hayfield, while they spent their down-time learning Gaelic from the two sisters at the farmhouse, and their evenings in traditional song and dance. A farm that lived by boat (the two sisters – Margaret and Janet MacDonald – were renowned competitive rowers), Dunnett saw this thriving island as 'a symbol of what we had hoped to find'. Leaving Calve, he wrote, 'what we did know, beyond all uncertainty, was that not this nor any other dawn would ever show us here a profitless land – a permanently beautiful but barren waste. It was on the contrary a rich land, if neglected; fertile in all but faith.'

When I stopped in Tobermory with copies of Dunnett's writing in hand, it was dispiriting to see that – despite the revival of recent years, and the many ways in which the canoe boys' desires for independent industriousness and local governance have been answered – Calve is unpopulated with no thriving farm and no little boats upon the water. Nothing could express more strongly the strange mix of fortunes these regions have faced since 1934: ever more appreciated and ever more viable in the age of instantaneous communication and devolved government, they remain fragile in the extreme and still far from achieving anything like the population and self-confidence of the era before clearances.

The Canoe Boys was the first major piece of kayaking literature in the English language; it is likely always to be the most significant. Dunnett was a major voice behind campaigns for devolution which would result in the foundation of the Holyrood parliament in 1999. Tragically, he had died in 1998. It was his kayaking odyssey that consolidated Dunnett's conviction not just that something must be done, but that real and radical change could be achieved. Few of the seasoned seafarers who looked at the canoe boys' boats in the autumn of 1934, shaking their grizzled heads and muttering 'it's too late in the year', could have imagined that one of the fresh-faced youths in front of them would reshape their nation in such a way. It's no surprise that all subsequent classics of English-language kayak literature, such as Brian Wilson's *Blazing Paddles: A Scottish Odyssey*, proudly wear the inspiration and influence of Dunnett and Adam on their sleeve.

Amid the depopulation the canoe boys drew attention to, there hid a tiny but intriguing counter-current. In the same decades that fertile isles were abandoned, some entirely unpromising skerries became sites of human occupation. As I moved from the Minch along Argyll's Atlantic coast I passed small, wild rocks such as Hyskeir, Skerryvore, Dubh Artach and Fladda whose lighthouses flashed greetings to passing ships. Other remote towers, such as Ruvaal (Islay) and the Irish isle of Rathlin, also threw beams across my path. On one particular night, sleeping beneath the Mull of Kintyre lighthouse, the vast array of sweeping beams became a trippy light show in which chattering snipe flew in and out of lances of illumination and Irish beams looked as bright as the light I slept beneath. In Ireland, the constellation of sea stars continued, from Fanad Head and Tory Island in Donegal, to the stupendous feats of ocean-bound construction at Bull Rock and Fastnet.

Modernity had bestowed every one of these unpromising sites with a tiny community for around a hundred years. These settlements were formed within the most dramatic feats of engineering in industrial Britain: the improbable accomplishments of engineers such as the Stevenson family who mounted lighthouse after lighthouse, each unique, on a vast range of skerries and clifftops.

Much has been written about these buildings. Yet those texts tend to place each light into categories that confine the diversity of lighthouse experience to something reductive and knowable. The division of 'rock light' and 'land light', for instance, has been taken to have a host of inevitable implications.[6] The typical 'rock light' was a high whitewashed building on a small skerry, manned by men whose families stayed ashore in small complexes of lighthouse cottages. For half the year these families would be split, so that keepers' children often failed to recognise returning fathers. In the typical land light, the family cottages stood in the lighthouse complex, so families stayed together and children were raised to polish lighthouse brass. Yet 'rock' and 'land' weren't the rigid categories they've been seen as, but extremes of a wide spectrum in which intermediate examples throw up surprises.

I visited several lights as I kayaked, exploring the unique set of buildings at each site and tramping through the remains of the keepers' improvised entertainments, including a tiny skerry golf course. But as atmospheric as any remote rock light was the structure on the island of Fladda (figure 8.1). On leaving Mull I paddled south into the Sound of Luing, where tides flushed the kayak into a wonderland of barren Atlantic rocks. And the first place the sea rivers spat me out was beneath Fladda's beautiful lighthouse. Although this light is not atop a classic pillar, elegantly tapered in a shape inspired by oak trees, and its rock is not far out in the ocean swell, it is a magical site to explore. Commodious buildings and a huge walled garden (green even in February) make the world inside the whitewashed walls feel secret and utopian, connected only aurally to the violent ocean world outside.

Despite their massive scale – sometimes thirty feet high by fifteen thick – there are intensely human details to the perimeter walls, including a peephole looking across the water to Belnahua: lighthouse inhabitants could keep up, in the usual Highland way, with the comings and goings of their nearest neighbours. But the racing tides, storm-prone waters and sea fogs ensured that through its short era of permanent habitation (1904–59) it was insulated a little from everyday interaction with the world beyond the seaweed boundary and operating according to unique rules dictated by the ocean.

The nearest inhabited island is Luing, and as soon as I landed there I sought information on the lighthouse community. Jane MacLachlan showed me the documents and pictures collected by the Isle of Luing History Group and told me family stories: her husband had been the Fladda boatman, charged with supply and service of the lighthouse, as had his father and grandfather. The tales, texts and photographs all undermined everything I expected of a skerry lighthouse. This was no cold site of monkish male celibacy but a place of warm family life where two households lived together and children played among the rocks. So many children were present that the two lighthouse families sometimes made up a population of nineteen. After seeing these sources I looked with fresh eyes at the guidance notes keepers at every lighthouse received from the Northern Lighthouse Board: there, staring out from the top of page two is an instruction for informing passing ships or watchers from the shore that the lighthouse required a midwife. I could find no way of fitting Fladda into the usual dichotomy of 'rock' and 'land' lights. It was a family lighthouse out at sea: an exception to the rules I thought I knew.

MacLachlan boatmen and Macaulay keepers seem often to have been awarded medals for bravery, but this was as likely to be for recovering children from sweeping tides after falls from Fladda's rocks, as for saving crews from ships such as the Latvian steamer wrecked here in 1936. With conventional schooling impossible, the children grew up helping to whitewash walls. This was a huge annual task in which immense pride was clearly taken: here, unlike at many onshore lighthouses such as the Mull of Kintyre, garden walls were kept as bright and perfect as the pillar itself.

The garden was Fladda's greatest resource. This is a rock without fresh water except supplies brought in small wooden barrels from Luing. It was once a rock with scant soil. But Luing and its neighbouring islands are rich with slate that was quarried for transport to Ireland; Irish ships reached Fladda before Luing itself and unloaded ballast at the lighthouse. Tons of Irish earth richer than anything on Argyll shores filled the lighthouse garden. Soon, Fladda was famed for its produce. Its large, fine carrots drew envious comments from the shore; home-grown strawberries became a highlight of lighthouse summers; eggs from Fladda hens made fine islet breakfasting.

Each spring and autumn the children played amid a strange menagerie of injured songbirds. In an era when naturalists habitually carried shotguns, keepers' families were unusually peaceable ornithologists:

rare migrating birds crashed into lighthouse glass so that dazed
survivors spent days hopping between the veg. Keepers described
the 'clunk, clunk, clunk' of redwings and wheatears hitting glass
during the biannual periods of avian chaos. Thousands of birds, one
remembered, could be found lying dead in the morning and their
bodies would be carted off by wheelbarrow. Other stories were
happier. Once, a gyrfalcon hung around for more than a week,
performing its acrobatics among the flurries of songbirds. On another
occasion, the keeper's son nursed a merlin back to health before
releasing it on the mainland. The first Scottish sightings of species
including the bluethroat were lighthouse casualties in this way. And
many island lights housed a library of some 400 books in which
subjects like history and folklore were swamped by guides for naturalists. Lighthouse children gained unusual coastal skill sets, able to
name species that even the most obsessive ornithologists on the mainland never had a hope of spotting.

After leaving Fladda I contacted Erin Farley, who was commissioned in 2013 to create an oral history of lighthouse keepers for the
Northern Lighthouse Board and the University of Edinburgh. A
masterful interviewer, the discussions she held are full of tales that
bring lighthouse worlds to life. Amid the descriptions of these
awesome engineering feats, I was struck by the ways in which lighthouses remained oceanic. The thickest of walls and wisest of
structural choices could never insulate interiors from the battling
elements. One keeper told how the carpets on lighthouse floors would
ripple and twitch when the gales arrived: they were sucked from
floors as weather transformed the internal dynamics of the tower.
Another told how decorating decisions were made by the winds'
intrusions. Ceilings were painted a dull dark brown because a paler
option would need repainting monthly: in certain winds the rooms
'could absolutely fill' with 'blowdowns' of blackening smoke. A third
keeper told of the impossibility of keeping dry when atmospheric

pressures changed. He described the 'water running doon' the interior of the building: condensation whenever a warm front came in. Oncoming waves could be 'like a tenement building coming towards ye', leaving the tower shuddering in their wake. And through all this elemental fury, fresh water was limited and electricity non-existent. Clothes would be smoothed (high standards expected) with a paraffin iron and food preserved in a pungent paraffin fridge which added its distinctive flavour to everything stored within.

Almost all interviewees agreed on the eventfulness of lighthouse life, one describing how her diaries of years at the light have the fullest pages of any in her life. Some kept themselves busy with crafts that were highly esteemed among keepers and often became competitive. Model boats and ships in bottles, made from driftwood collected after storms, were particularly well regarded. Feathers of migrant birds were turned into fishing flies for immediate use or for sale onshore. Those with particular skill in any craft gained wide renown, such as Andy Flaws, who could turn any piece of wood into a beautiful model trawler, precisely recreating, from memory, elegant boat styles such as Scottish east-coast Fifies or Irish Galway hookers. Almost everywhere, keepers gardened against the ocean, running local seaweed through imported soils as they tried to generate vegetation so unlikely that the eyes of other keepers would widen in wonder. Such glories weren't universally achieved, however. One keeper at the exposed Irish light of Slyne Head noted that 'we existed on tinned foods. The advent of the fridge was the greatest thing that ever happened.'

Time and again, the conversations families had with Erin Farley returned to the importance of the lighthouse children:

> If I hadn't had my wee lad I think it would have been very very difficult . . . I think if you talked to any of the women they'd say it was the children . . . they were the glue, you know . . . I just had the one but there was big families.

One boy, Alistair Laws, had a mother posted to a shore light and a father posted to the nearest offshore site. Even though this was one site from which children were officially excluded, he could hitch a ride out and stay on his father's island for weeks. He remembered the site as wide and sparse, but with an extraordinary amount to see: nature, weather and the activities of boats and keepers. He recalled the sudden descent of thick fogs and the excitement of hearing the twelve foghorns that would 'just blast your ears off'.

In all these ways lighthouse keepers improvised ways of life that operated in contest and collaboration with the powers of the sea and that offer unique insights into the possibilities for human inter-action with the coastline. Yet the experience of these families is also instructive in relation to the cultural diversity of these regions. Keepers were moved from post to post, sometimes lasting less than a year at a single light, occasionally more than twelve. East-coast and west-coast keepers were thrust into close proximity, forced to attempt to comprehend each other's strange vocabularies and stranger habits. Some keepers were middle-class people who'd enjoyed years of global travel before deciding lighthouse-keeping would be a suitably bohemian vocation; others were merchant-navy boys or workers in industries such as fishing, forced to look for alternative employment during slumps. Such differences were often expressed through food:

> You were in wi many cultures, like I was wi a Fifer, a Hebridean guy from Barra, and a Glasgow man, and a guy fae Arbroath. So it wis aa different cultures . . . The Eastie Coasties, we like mince and tatties, white pudding, skirlie. So we tried to introduce that and the ithers went, 'ugh!' And the Glasgow guy, he couldna cook to save his live, he was the Principal, and aa he did was open tins . . . He never tried to fry an egg or anything cause it was above him . . . or below him, with the way ye look at it.

For a west coaster, being posted to Orkney or Shetland was like going to 'Timbuctoo'. It involved a kind of isolation that most had not experienced, so that ending up, later, on an Islay light 'with all the shops' felt positively metropolitan. An east-coast keeper recounts his problems adapting to the 'weird and wonderful' habits on Lewis, where herring were plucked from barrels with rusty tongs.

And lighthouse families' interactions with locals varied wildly. Many found it easy to integrate briefly, recalling that they weren't seen as intruders or interlopers, because they'd been sent to do an honourable job. Others confronted problems, especially where language or religion proved a barrier. The Northern Lighthouse Board's identity was closely tied to the Church of Scotland, meaning that in Catholic regions such as Barra or Uist, keepers were distinctly foreign to the locals. English-speaking keepers in Gaelic areas could also struggle, and the board worked hard to ensure that challenging linguistic combinations were avoided. The combination they worked hardest to avoid was one English-speaker posted with two Gaels at a light.

Although the Irish and Scottish lighthouse services were distinct, their experiences were intertwined. The same ships, travelling perhaps between Limerick and Glasgow, would be reported to Lloyd's of London by telegram from Irish lights moments before Scottish lights would do the same. The cultural challenges of lighthouse-keeping are most widely attested in relation to Irish lights. Keepers posted to Donegal sites such as Fanad Head and Tory Island brought families into areas where the only schools were Irish-speaking, while Irish-speaking families would regularly be posted beyond the Gaeltacht. Children's experiences of moving between the diverse coastal worlds of the Irish west could thus be constantly disorienting.

Stories of lighthouse rocks are intriguing in themselves, yet they also pose a larger conundrum. The easiest narrative to explain the emptying of islands in the face of industrial modernity is that the risks of sea travel became unacceptable in a society where safer, if not faster, alternatives became available. Yet the abandonment of coastal travel and the movement of trade inland occurred just as the engineers of lighthouses did much to tame the coast's threat. The delicate and hyper-specialised economies of industrial modernity were built upon predictability and the mastery of weatherproofed land travel which removed wayward coastal waters from the equations of trade. The sea became a space of large-scale voyages in ocean-conquering megaships, and less a medium of everyday circulation. This ultra-modern need to control risk dictated the new geographies of Britain and tethered power and wealth ever more decisively into metropolitan orbits. The lighthouse-builders' heroic risk management was but one symptom of the new outlooks through which fresh forms of thalassophobia entered western life.

After reaching Kintyre I restarted my journey from the Irish side of the sliver of water that separates Argyll from Ulster. I walked up Ireland's northernmost headland, Malin Head, before I rounded it in the kayak. The view was Hebrides and the hills of Lowland Scotland. This was just a slightly different vista on the worlds I'd been moving through for weeks.

The historic significance of this north-west coast of Ireland has been even more unhelpfully underestimated than the Scottish seaboard. Histories of Ireland at sea were long preoccupied with the wealthy ports of the east and south, through which grain was traded and naval power wielded. Documents like state papers encouraged that trend: it was along the seaboards controlled by the Normans and the English, not by those oriented north, that official records proliferated. As Colin Breen, a historian of the Irish coastline, has noted, 'absence of commentary on the west might suggest to the casual reader that it was largely uninhabited or underdeveloped'.[7] Similarly, hosts of books treated Anglo-Irish links as well as Anglo-Scottish ties, but until recently many fewer have explored the far stronger connections between Ireland and Scotland.

Only in recent decades have historians and archaeologists begun to acknowledge that the power, wealth and sophistication of the west coast simply belonged to a different world than south-east Ireland did. Archaeological sites in the south-west reveal more Iberian trade than English. In Donegal webs of Atlantic exchange run northwards. The intellectual innovations of early-medieval Ireland were Donegal developments led by figures such as Colmcille (St Columba) and his later follower Adomnan. It was the Irish clerics they inspired who spiralled outwards on vast oceanic migrations, following geese to Iceland or skirting Scandinavian fjords.

Later, as outsiders exerted increasing control across south-east Ireland, the west conducted its own flourishing seaboard trade, exchanging fish for wine and drawing large levies from visiting fleets. But the quays and ports of the west are tiny, and the oldest structures are buried in newer developments that use the same tides and channels, so archaeologists long learned little of west-coast trade. Fifteenth-century fleets, 600 strong, paid handsomely for access not to grand stone quays and sprawling wooden jetties but to the awesome natural anchorages amid the ragged coast's complexities. West-coast settlements have now yielded to excavators a wealth of Valencian lustre vessels, Saintonge ware from western France, south Dutch maiolica and even Chinese Ming, showing wealth and global reach long before the seventeenth century when English wares begin to proliferate.

More than any other place on my journey, Ireland and Irish innovations were lynchpins of Atlantic worlds bigger than those my kayak could reach. The Irish brought modernity to the water: it was an Irish master mariner, Robert Halpin, who created the modern global village of instantaneous communication by laying 26,000 miles of ocean cable that 'tied up the world'; another, Francis Beaufort, gave the world the scale of windspeeds crucial to modern seafarers. They were inheritors of a vast tradition. Ancient Irish kings and corsairs colonised Brittany, died at the mouth of the Loire or the foot of the Alps, and made peace or war with the kingdoms of Iberia and beyond. Irish holy men, tracing a 'stepping stone' route via the Western Isles and Iceland, may have been visiting the Americas a millennium before Columbus. Indeed, the sanctity of the early Irish saints was judged in seafaring terms: those who possessed the maritime skill to cross the fiercest waters were the strongest in faith. Narratives of saintly seafaring were penned by sailors and produced for readers who were clearly accustomed to the perils and joys of ocean travel. There are tales throughout the Atlantic and Mediterranean worlds that associate Irishness with sanctity and boatbuilding. A traveller on the Tyrrhenian Sea, for instance, might happen upon a small chapel to San Pellegrino in the Tuscan town of Lucca. Pellegrino was an Irishman, of perhaps the eighth century, who settled at the foot of the Apennines after pilgrimage to Palestine. His chief miracle is boatbuilding: when he cast his cloak onto the water it became a boat, and his staff, thrown after it, became the mast.

As early as the sixth century, texts recording such stories and the beliefs they expressed were written not just in Latin but in the Irish language. This tongue spread north along Atlantic coasts, creating an Irish world across 1,000 miles of seaboard. Only slowly, and only partially, did Scottish Gaelic cut loose from Irish origins; indeed, where it was once assumed that nations were defined by language (that is to say a border between Scotland and Ireland was defined by the speaking of Scottish and Irish on either side) the reverse is really true: the idea of the nation led gradually to the fiction of two languages, consolidated from the many diverse Gaelics that were spoken. Only after the fact, in the nineteenth century, did the fiction gain a semblance of truth.

In some periods of history, formations such as Dalriada have straddled these sea zones and even when the patterns of power didn't create connections, trade and culture did. The very name 'Argyll' is often glossed as the borderland dividing the Gaels, but should perhaps be interpreted instead as the bridge binding Gaeldom. Later, labour circulated freely between these regions and Scottish fishermen who landed at Donegal ports found they could communicate – slowly and stiltedly – across Gaelic's divergent branches. Seasonal trade still circulates thousands of workers between Argyll and Irish farms, fisheries and construction. I learned while travelling that teenagers from Argyll islands such as Islay still hitch lifts on fishing boats to party in Donegal and Belfast. There seemed always to be fellow feeling between people I met in Argyll and Donegal. Irish islanders habitually referred to the Scottish language as Scots Irish; one Islay-born ex-fisherman I met in the Donegal town of Dungloe said that the islanders on each side of the border look across at each other and think 'we're you, really, or we'd have been you with one more defeat, or one defeat less'; many expressed great puzzlement that more Scots hadn't embraced the opportunity for independence in 2014.

More than any other regions I travelled, the distinctions between land and sea mean little here. Historic Irish societies blurred all distinctions between surf and turf. Their power lay in ships as well as heifers and horses. Fishing took place in products of the pastures: ox hides were stretched, stitched and tarred over wooden frames to make their characteristic boat, the currach. Small and light, capable of extraordinary buoyancy in surf and easily patched up when shredded on rocks, these boats seem adapted for inshore waters: a rough-coast vessel characterised by cheapness and expendability. Yet, in their northwards travels, these are the ships that populated Iceland. A fourth-century text, *Ora Maritima*, by the Roman poet Rufus Avenius described Irish sailors, whose 'boats sail freely on this rough expanse of sea'. 'Amazingly', Avenius continues, 'they contrive their ships by stitching skins together, and cross the sea in open leather.' A century later, there were rumours of Irish skin boats attacking Roman Britain, their oars churning the water white; Lucan mentioned their raiding voyages across 'the swelling main' to Gaul, in boats made from 'willow boughs . . . and slaughtered kine'. Here was an entirely different answer to the challenges of tree-scarce coasts than that Shetlanders found in links with wooded Norway.

While currachs were made from the fauna of the rough land where livestock grazed, lobster pots came from its flora, handmade from heather, and ropes and lines were donkey and horsehair. And the boundaries of land and sea were also muddied in reverse: when mortar was used to cement stone houses, this was not ground from limestone as elsewhere but from limpet shells, bringing the sea into the hearth of the cattle farm or clachan (village). The fabric used for threshing grain was worn-out sail. One evening I was camped on rough peaty bog, reading about Irish coastal life, when I found myself staring into a dark, clear pool of bog water, seemingly unfathomable in depth and edged with fronds and foliage. It was a perfect tide-line rockpool in photographic negative. These worlds of plough and net cannot be disentangled.

Today, the Irish language entirely predominates on many islands and peninsulas: as in Wales and Scotland, linguistic richness is synonymous with the coast. The Irish Gaeltacht includes no large islands like Lewis or Skye: instead, small ocean jewels are liberally scattered down the coastline and of these around half retain the language to a degree rivalled in Scotland only by Barra and Lewis. Many of these followed trajectories entirely the opposite of the Irish mainland: the population of Inis Mór grew through the great famine, while the Aran Islands (Oileáin Árann) were untouched by potato blight and, according to contemporary witnesses, untouched also by destitution in Ireland's darkest years. Yet many have historically been so detached from elsewhere that a single visitor carrying measles could spell demographic disaster.

Large tracts of coastal mainland, such as Connemara and west Cork, are also islands of Gaelic in the inland sea of English. Irish Gaelic's relative strength lies in the fact that, despite massive depopulation and the irredeemable damage done to the language by the 1840s famine, the Atlantic edge remains more populous than the western coasts of Scotland's Gaelic heartlands: islands such as Arranmore, with 650 residents on just 6.5 square kilometres, have no Scots Gaelic parallel.

Indeed, along all these coasts, a visitor from Britain must rethink assumptions about settlement. Large houses and single-track roads line headlands that would otherwise feel remote. Almost imperceptibly at first, they thicken towards village centres. Amid this flux of varied populousness, there's no logic to the idea of a contrast between 'urban' and 'rural'. In ways I found surprising and almost surreal, tiny islands like Inishcoo host the grandest of eighteenth-century town houses with deadly tide-races feet below their doors and cattle swimming between sea rocks at slack tide; massive swells and breakers storm back gardens in ways I've seen nowhere else; shops are found in spots without passing cars but from which webs of well-used waterways sweep out. The ideal shop was a building like the old Corn Store in Killybegs, Donegal, with a frontage on both road and water.

Where, on the whole west side of Lewis there's a single pub constantly in danger of closure and facing away from the sea, a similar distance of Donegal coast holds hundreds of watering holes looking straight onto the ocean. It is only those who travel western

Ireland inland who could ever imagine it characterised by lonely loughs and desolate hillsides. Eighteenth- and nineteenth-century maps make these regions look deserted: they show few roads in coastal regions. Yet there was widespread human movement everywhere. The tradition of conducting business by small boat was far more developed, and has been far more resilient, here than on any other Atlantic coasts.

Such shorelines do modernity differently from landlocked approaches to twenty-first-century life, their viability aided by the uniquely strong roles the Atlantic continues to play in Irish consciousness. Even for many Dubliners, what it means to be Irish is inflected by the idea of a nation physically and culturally focused on the scar tissues of its ocean fringe. Perhaps this is one reason why Irish island life is centrally subsidised (though extremely unevenly and often inadequately) in ways the Scottish isles are not. The association of Ireland with the Atlantic has also given rise to a host of strangely persistent fictions, such as the idea that the Dingle Peninsula in County Kerry is the westernmost point in Europe (a whole European nation – Iceland – lies further west than this), or that the sea cliffs of Slieve league are the highest in Europe (the cliffs of Faroe are far higher). The persistence of these misconceptions illustrates, perhaps, the power of the urge to show the world that Ireland is the capital of the North Atlantic Ocean.

But this urge is wholly justified, even when some of the claims made in its name are not: it has caused substantive revisions of history over recent decades. Regions of the west, such as Connemara, were once thought to have been disconnected and sparsely peopled till invaders such as Cromwell forced half of Ireland into them. But archaeological evidence is transforming that picture: from the Neolithic to the Middle Ages, Connemara, a region of a thousand sheltered quays, was a hub of large-scale trade and travel. That the myth of remoteness could survive so long is a measure of how far historians, if not the people of Connemara themselves, have lost their understanding of the sea.

In film, activism and scholarship pressures now exist to re-evaluate Ireland's coastal heritage in order to strengthen commitment to a maritime future. As elsewhere on the Atlantic edge this urge began to be felt strongly in the 1970s. An extraordinary transatlantic voyage in a fragile cow-hide vessel, undertaken in 1976 by the author and adventurer Tim Severin, was an early example of these impulses. In recreating the historical voyage of St Brendan, Severin revived an ancient tradition of ambitious and technologically advanced seafaring whose veracity had been widely doubted. He recounted that journey as *The Brendan Voyage* (1978): one of the great examples of Atlantic literature. Like the historical society of Ness on Lewis, this was an effort to re-narrate the Irish coastal past, showing that traditional did not mean parochial or backward and that the ocean had been connective tissue between distant historic cultures. Severin was an activist for an outward-looking, oceanic Ireland in an era when Irish politics was focused on farmland.

Five years later, a week-long convocation of fishermen, politicians and scholars met in County Clare to discuss the neglect of Ireland's maritime heritage and to consider strategies for its reclamation. They defined the problem through fusions of economic and cultural rhetoric:

> For too long this nation has been kept ignorant of its maritime heritage. For too long this state has neglected the seas that forever lap its shores, betraying the efforts and sacrifices of countless generations of Irish navigators, sea-fishermen ship-wrights, the very origin of human life on our island. Ireland is the poorer for this wanton neglect, for which the political leaders of all parties and the national education system they have fostered stand indicted. But it will not always be so. Irish

people will eventually move to enter into their priceless maritime heritage.[8]

Another contributor insisted that the Irish were an Atlantic people 'looking south and west', whose 'links with North America are stronger than those of any comparable European nation'. He labelled the Irish and their Atlantic doubly unique: the sea was exceptional in the range of 'aspects of Ireland' it encapsulated yet never can 'an island people have so ignored the impact of the sea on their existence'. This marked the beginning of a vast but stuttering renaissance, evident throughout my Irish journey. At every stage I'd meet people whose lives have been devoted to battling that neglect and I'd witness a multitude of strategies by which the Atlantic has been reintroduced not just into the cultural life but the political consciousness of a nation.

At the mythological beginnings of Irish history, islands were the engines of change. When the godlike race of Tuatha Dé Danann conquered the mainland, they banished the previous inhabitants, the Fir Bolgs, to scraps of rock in the Atlantic. In doing this, they unleashed another force: those islands were peopled by an uncouth but powerful race called Fomorians who, under their king Balor, god of darkness, began to spread havoc on the mainland. Only when golden Lua – half Fomorian, half Tuatha Dé Danann – sought the aid of the sea god, Manannán Mac Lir, was Balor struck down and Ireland and its islands brought peaceably together. In myths featuring Manannàn Mac Lir, his ocean realm is a place where currach races chariot: the Atlantic is a mirror of Ireland containing parallels to every aspect of social organisation. Only limits to human perception make ocean seem less hospitable than the land.

My Irish journey began with a voyage to the most mythic island of all. Nine miles offshore, across some of the fiercest seas in Ireland, Oileán Thoraí (Tory Island) is a truly spectacular ocean rock that was Balor's home: his fortress of darkness and disobedience (figure 8.2). So separate is Thoraí from the mainland that major factors in Irish history, such as potato blight, never touched its shores: in the aftermath of the most profound demographic catastrophe in modern European history, the great famine, the population of the mainland overlooking Thoraí was reduced by 41.5 per cent but Thoraí's only by 16 per cent. I spent the night high on a cliff-bound promontory, known as Dun Balor, where the dread king is said to have held court and where, it's rumoured, he buried nine tons of gold. Sure enough, four ranks of ancient fortification are evident in the earth, with circular hut circles behind them: Balor and his gold may be mythic, but a stupendous crag-bound stronghold attests to Thoraí's Iron Age power. The red granite cliffs of the island's thundering north lean into overhangs, protrude into towering stacks and recede into spume-filled caves and arches: some of the most thrilling kayaking in Ireland. The land behind sweeps gently down to small low fields at the southern shore which were, for centuries, extensively farmed. When great storms approach from the north it's said the sea spills over 'the cliff edge of Europe' and pours in torrents down the island incline, washing away crops and even livestock. This is often interpreted as Thoraí's treacherous crags invading its cherishing south: during storms, Balor's warlike memory invades the quiet hearth-lands of Colmcille. Yet in truth the cliffs are protective of the lowlands and, in their own ways, intensely nurturing. When I arrived, late in March, spring was conquering the island cliffs first: they were eruptions of bright white flowers of scurvy grass while the southern fields were barely throwing off their winter brown. Before the winter geese had left the low fields, razorbills and guillemots were perched on ledges, reshelved like a library of black-and-white books.

Each feature of Thoraí's contrasting landscapes is named in detail, the north largely for myth or topography (such as Morard, 'the Great Height'), the south for culture and history (such as Pairc an Lin, 'the Field of Flax', or Feadan an Wasp where, in 1884, HMS *Wasp* was wrecked while attempting to collect taxes). But these names are still more reliant on oral custom than are Scottish island sites. The gulf between detailed local knowledge and the sparsely marked leisure maps carried by travellers is far greater than that in Scotland, where general-use maps are among the most detailed and multilingual in the world. The weakness or strength of oral custom, depending on your perspective, is that there are as many different versions of island geography as there are islanders; great arguments can sometimes take place, one islander told me, over which precise patch of water was the setting for any one of Thoraí's historic sea battles.

As I set up camp among the fulmars on a fearsome overhang marked with roots of past walls, which suggests an Iron Age lookout point, I had the first of many serendipitous meetings that seem to happen frequently on Irish islands. The island nurse wandered across the clifftops, stopped to comment on the exceptionally early arrival of the puffins, and took me to meet Thoraí's king. This island title is said to have been established by Colmcille in the sixth century and conferred on a man named Duggan who was the first proud pagan islander the saint converted. The king spoke of Duggan and Colmcille, whose monastery stood in the middle of Thoraí's west town, as if they belonged to living memory. Much later, I stumbled back from the island social club to my clifftop roost, having talked of Thoraí life and heard the barman sing island songs till the morning's early hours.

In newspaper cuttings framed behind the social-club bar, and in every word uttered by the island king there was evidence of Thoraí's phenomenal sense of independence from the era of Duggan onwards. This rock has been owned by landlords from Glasgow, Manchester and Birmingham, yet all the wealthy men who sank vast sums in its purchase found their plans and profits unravelled by island spirit. The women of Thoraí gained a particular reputation for strength and rugged defence of island ways that is worthy of Balor himself. They had most to lose from the loss of their culture since Thoraí women, not just eldest sons as in the rest of Ireland, inherited land and wealth. Between 1872 and 1903 not a penny in rent was paid to Thoraí's owner, and the courts in Liverpool, who ruled repeatedly against the 400 islanders, flapped in consternation when they found themselves powerless to inflict censure. One cutting framed behind the bar recorded the views of the landlord's agent, Colonel Irvine, in 1895 that 'the island was in a state of absolute lawlessness, and the life of anyone going to collect rents was in the utmost peril'. He had, the *Londonderry Sentinel* noted,

> endeavoured to get there under the cover of a picnic party, but the natives, learning that he was on board, absolutely refused to allow them to land. A tax-collector who went there was set upon, shipped, and set adrift in a boat, and another was assaulted, and returned much less of a man than he went there. The inhabitants were principally women, and they were very lawless.

The previous landlord, a Glasgow industrialist, had set out on his time as island-owner full of bright-eyed optimism, earnest in his calling to enlighten benighted tenants with the benefits of improved agriculture and the 'Protestant ethic' of honest labour. All over Ireland, small arable patches, circulated annually between villagers, had already been rationalised into square fields farmed for commercial acquisition, not community subsistence. Until this time, the Irish landscape's structure had been made up of infield crops, outfield grazing and the small communal smatterings of homes – the clachans – which looked, in the words of the folklorist Estyn Evans, like they had fallen 'in a shower from the sky'.[9] These homes were now replaced by farmhouses located in large private fields.

A momentous political undertow drove this change. Enlightenment progress involved the universal integration of the monetary economy and the elevation of wage labour as the only acceptable form of production. Attaining these twin goals meant wiping out ancient common rights and imposing new systems in which individual property and measurable order prevailed. The lock and key, or the padlocked gate, symbolised a new commercial logic, incommensurable with the moral economy of rural life in which ownership and private property had been a less significant ideal. From the late sixteenth century, Irish rebellions against English overlords were followed by surveys and the making of maps that detailed in English the baronies of Gaelic lords and their value as potential property of the English Crown; this had been an early step in the gearing of Irish shores to the needs of distant commercial interests. From the late eighteenth century, the form and ownership of the Irish coast was mapped with new scientific precision by chain-dragging soldiers and subjected, by ever more powerful centralised authorities, to intensive statistical analysis.

The ideology of progress involved annihilating the communal culture of the clachan not because it was anachronistic or failing but because it represented an alternative set of values that could threaten, through competition, the fragile ideals of the new political economy. In clachans such as those on Thoraí, closely packed dwellings facilitated rituals of shared labour and leisure in which the binaries that structure modern urban life made little sense. The stark divide between work and recreation, for instance, was meaningless when work was the occasion for song and story and, as on Rob Donn's hillsides, the places of duty and escape were the same. Mingled functions of time (work/play) and space (workplace/home) have been known by the term 'through-otherness', an idea still used by scholars and present in the poetry of Seamus Heaney. To the eyes of visitors from cities, through-otherness often looked like mere confusion (in modern Ulster dialect it has gradually taken on this meaning, coming to imply simply messiness). But the communal structure of the clachan, alongside the low-intensity nature of potato farming, and the extent to which rough weather dictated downtime, shaped a rich culture that regularly impressed those few visitors who were not invested in an evangelical mission on behalf of commerce. Alexis de Tocqueville, for instance, noted 'stunningly vigorous and civil social cohesion, amid the mud and rags'. The postcolonial critic David Lloyd argues that English culture worked so hard to demonise the idle Irish in this period because their health and flourishing amid material deprivation revealed that 'bare subsistence' can sustain human life: their 'excess of lack' was a dangerous reminder that the commercial quest for financial accumulation was a choice not a necessity.[10]

In the face of island resistance, Thoraí's owner found his vision of progress hard to implement. He met islanders and island culture that were strong and single-minded. Demands for change were ignored, beside the small concession of enlarging, slightly, the scattered strips of existing infields. Thoraí powerfully demonstrates how much was at stake – the imposition of an entirely new social order – in the reorganised landscape of the nineteenth and twentieth centuries. This was just one flashpoint amid centuries of examples of Thoraí independence, even including refusal to accept the innovation of daylight saving time. No other island I visited proved so consistently resilient to the forces of centralisation. When I asked the island king why Thoraí had proved so gloriously independent, he echoed the sentiments of the *Londonderry Sentinel* and the findings of historians, such as Lynn Abrams, who have shown that island cultures were occasionally able to sidestep elements of modern patriarchy. 'Our strong women,' was all the king said.

Thoraí's advantageous distance from the centralising forces of land is matched by its centrality to the traffics of ocean. It has proved a crucial landing spot or stepping stone to countless exiled or defeated movements: Scottish Presbyterians, for instance, fled here in the 'killing times' after the failure of the Covenanter rising in 1679; folk memory holds that, as well as the boat of emigrants who integrated into island life, a second vessel, in which all aboard had perished of exposure, was washed to the bottom of the island cliffs. For covenanters, as for many others, Thoraí offered a distant and dangerous dream of tide-bound escape from authority.

Looking back through statistics of island production, several implications of this oceanic existence are evident. In the early nineteenth century, islanders did spectacularly well from the kelp boom, particularly because isolation allowed them to resist the rising rents imposed by landholders on more proximate tenants. The island in this era was filled with horses: the only beast sturdy enough to pull kelp sleds on iron runners. The kelp boom represented, however, an unusual turn towards the ocean for a society whose economy had, surprisingly, been centred on its arable and pastoral land. Until this time, agriculture, not shoreline trades, had been the island's economic mainstay and donkeys, rather than horses, the islanders' chief vehicle. Working days were long and close to the soil, inflected by endless vagaries of weather, which in an hour could spoil three months' hard labour. The multitude of infield strips, whose patterns can still be traced across the ground, were used for potato, rye, barley and oats. After the collapse of kelp it was the combination of the produce of the land with Thoraí's place on routes of the new steamships that defined island life, and the shifting balance between the various arable crops hints towards one secret of Thoraí's fierce independence. As the number of passing steamships increased, so did the proportion of Thoraí given over to barley. Despite the fact that weather-resistant oats were the resource best suited to the island, and that rye was uniquely useful for its dual purpose as food and thatch, three times as much barley was soon grown as either crop. Official documentation lists this as animal feed, but never have hens had appetites so prodigious that these figures could make sense.

The peat bogs, among the island's most significant resources, also underwent a strange transformation at this moment. Early descriptions of island life describe the cutting of 'chocolate rectangles' with 'gleaming loys', whose burning joined the aromatic ocean in flavouring Thoraí air. Without the fragrant, intensely calorific turf that held long winter nights at bay on coalless, treeless islands, population histories would be the weakest echo of what they are; and this peat is yet another oceanic gift, its conquest of swathes of land a result of damp Atlantic air. But today, it is the mingled smoke of wood, coal and only imported peat that rises from chimneys to tussle with the smells of washed-up wood and weed to ensure that recollections of the island are always olfactory. Just as barley production boomed, the island's bogs were destroyed: peat was stripped to the point that all fuel had to be brought in from elsewhere. Never could clachan hearths have guzzled energy on the scale required to achieve that disaster.

The explanation for this dual mystery is that islanders had found a potent supplement to their limited incomes: the production of vast quantities of *poitín* (home-made spirit). This phenomenon is the most dramatic symbol of Thoraí's oceanic status: taking advantage of the ready market provided by passing steamships, the island had become the moonshine capital of both sides of the North Atlantic. Where other islands saw regular visits from priests and constables who seized and destroyed their stills, Thoraí's distance offshore kept such authorities at bay or gave islanders ample warning of their coming. This lucrative and honest (though far from legal) trade was one among many reasons why Thoraí folk wouldn't surrender separateness lightly. The island had found its own route into the world of integrated commerce.

Remarkably, it was only in the twentieth century, when bare bogs limited *poitín* production, that the sea itself became the source of Thoraí's sustenance. Fishing was a phenomenally small-scale affair. The island's cow-hide currachs were tiny (far smaller than the longer currachs favoured by other island cultures) and used for ventures close to home. Nineteenth-century visitors praise their ability to land through heavy surf, but bemoan their total lack of comfort, without so much as a plank for a seat. Thoraí's cliffs are scattered with cairns: these were markers for islanders onshore to find sight lines to favoured fishing grounds. The clifftop watchers had gannet's eye views of shoals and could signal the currachs below towards silver suppers. It's still possible to stand on the crags and identify the spots where past islanders expected to see the ocean's harvest gathered.

In 1903, the Congested Districts Board (a body described by a historian of the Donegal fishing industry as 'the most significant engine of progress ever to be established in the west of Ireland') secured permission from the islanders to acquire Thoraí from its beleaguered landlord.[11] Aware that herring in the region were not being 'properly exploited', they began to invest in a Thoraí fishing fleet. This was to be one component in the grand machine of their dreams: an organised fishing industry along the whole west coast, connected to Belfast and Dublin by a modern rail network. The board drew in expertise from Shetland and Ullapool. They built a proper harbour and supplied large wooden yachts with modern tackle. They even bought fish directly from the fishermen, aware that it was lack of rapid access to markets that most reduced islanders' prospects of success.

Between 1895 and 1914, the Congested Districts Board supplied almost 200 Scottish Zulu-class sailing boats, free of charge, to Donegal fishermen, including Thoraí islanders. These dramatically beautiful boats are instantly recognised by steeply sloped sterns and huge lug sails (four-cornered but far from square, extending on both sides of the mast and permitting prodigious speed). Such apparent eccentricities were ploys that made Zulus manoeuvrable despite their unhelpfully short keels: they were designed at a time when each boat's harbour dues were calculated by keel length. Thoraí's customary division into clanns – in which extended families formed units who worked land and sea together – was now used to form Zulu crews.

By the time building and buying projects came to fruition local herring stocks were in precipitous and terminal decline, while large Scottish steamers had begun to work Donegal waters on a scale beyond the scope of sail. The brief but significant heyday of the Zulus is now more or less forgotten, its stories tainted by deflation born of failure (I found it only in books, not mentioned once in conversation with people I met). For the rest of the century, lobsters were Thoraí's main resource and a pattern developed of half a clann spending winters labouring in Glasgow or Argyll and summers creeling rough Thoraí waters.

Throughout these changes, the weather remained the greatest threat to year-round occupation of the island. The nine-mile sound between Thoraí and the mainland is a funnel for rough seas, aggravating swell from molehills into mountains. My own crossing had been suitably dramatic: far rougher than I'd guessed from shore and providing the kind of cumulative climb and descent that could equal a day's hillwalk (figure 8.3). But this was a calm day. In 1984 storm conditions held for eight weeks, isolating the island entirely. Thoraí at this time was still without metalled roads, piped water, waste collection, sewage treatment, medical provision or a reliable electricity generator. The ordeal was such that twenty-two island families applied for housing on the mainland and ten had soon moved. Now that peat cover was no more, fishing in decline and agriculture on the island all but ended (even such island staples as potatoes and milk were imported) the pressure intensified for a total emptying of Thoraí.

Three things neutralised that pressure, perpetuating islanders' powers of resistance and sense of self. It is symptomatic of the current trajectories of Atlantic islands that all three were cultural developments, not related primarily to economic productivity. The most immediately evident of these factors is the rising support for the Irish language. The strength of Irish here (not to mention the unique Thoraí modes of its utterance) brings Gaeltacht funding for infrastructure such as the large new health centre on the foreshore.

The other two factors in Thoraí's resilience are both more emphatically artistic. Thoraí is a last stronghold, along with Connemara, of *sean-nós*: a song style that is the product of a highly integrated community's response to a centuries-long way of life shaped by the hardships of Atlantic living. Sandwiched between the house of the king and the true ruler of Thoraí, the ocean, the social club into which I walked during my night on Thoraí is a sacred space of island tradition. As the ethnographer of island song, Lillis Ó Laoire, has made clear, this is not a community that sees entertainment as separate from cultural formation or artistic craft as distinct from communal organisation and the creation and maintenance of social ties.[12]

But the most distinctive source of resistance to the island's emptying is the tradition of art around which much island identity now revolves. In 1954, the painter Derek Hill began visiting Thoraí. He occupied an old signal hut at the west end of the island's cliffs. Islanders refused to sell it to him, though he proved as stubborn as them and was soon told 'you can rent it till you die'. The king lent me the key to explore Hill's hut. Ten feet by nine square it sits by a small brackish dew pond that Hill used for washing; within, Hill collected ocean artefacts such as the vertebra of a sperm whale. He cooked using seawater, doing innovative things with pasta, cabbage and seaweed, often picked not from the shore but from the hut's roof after storms. Rising early, he painted in the dawn glow, then rested till the ravishing evening light and painted again. Years later, Seamus Heaney invited Hill for dinner and recalled how the ageing, wheelchair-bound artist found the sight of the sun going down so unbearable that he asked to be moved to where his back would face the window. Hill's best days on Thoraí were those when weather was so rough that he worked on his belly beside the hut, painting spray and spume with gulls' feathers salvaged from the rocks. At least one canvas was lost to the elements. Derek brought many luminaries of the artistic (anti-)establishment to this hut, including the Marxist art critic John Berger who described the island houses of west town 'cluster[ed] together like survivors huddled in the stern of an open boat in a heavy sea'; this was before the current electricity generators were installed, and Berger recorded that 'the rooms in which the islanders live are dark as hutches'.

The true significance of Hill's three Thoraí decades, however, arrived when an islander named James Dixon, who had been secretly painting at home, looked over Hill's shoulder at his work and insisted he could do better. Hill gave him paints and canvas (though Dixon refused brushes, saying he cut his donkey's tail three times a year). Dixon painted Thoraí shores and seas, using island memory – notable wrecks and famous catches – as his theme. In Dixon's words, this was a historical record of brutal times and places: 'nothing romantic . . . little boats fighting and clashing with waves and winds'. He adopted a strange perspective, seeing the world as if through a seabird's eyes, but with key features such as ships and houses stretched side-on; his work therefore evokes something of medieval cartography.

Hill and Dixon encouraged other island painters. The first off-island exhibition of Thoraí artists was held in Belfast in 1960, while Dixon's first solo show took place in Dublin seven years later. In subsequent decades exhibitions toured Europe and crossed the 'Thoraí pond' to New York. The island king at the time of my visit, though he sadly passed away soon after, was Patsaí Dan Mag Ruaidhrí who had painted Thoraí land and sea for half a century. His stylised skylines capture the charisma of the Donegal hills, seen from the vantage of his house above the harbour (the largest council house, he often said, in Ireland). He pulled me up for asking whether Hill had given islanders artistic training (training, after all, would never have sat well with headstrong island independence) and insisted that encouragement and generosity were the right ways to think of Hill's role. The barman singing on the Saturday night I spent on Thoraí and the nurse who introduced me to the island are both also painters whose boats and seascapes hang on island walls. It's striking, though perhaps unsurprising, that when island productivity has otherwise collapsed to record lows it's the arts, and public fascination with the imaginative interpretation of Atlantic coastal landscapes, that have allowed island life to persist. This art is resolutely true to Thoraí's independent nature. Two panels painted for the chapel by Anton Meenan (the first island artist to acquire formal training) show St Colmcille and Balor side by side: the twin muses of this Janus-faced Atlantic outpost.

Having survived the privations of the 1980s, aided by public figures such as Derek Hill, the last twenty years have brought expansion of Thoraí services, from a new secondary school and a regular ferry with room for seventy passengers. The role of the king, once focused on governing the rotation of land and shore resources, has been consolidated in fighting the island's cause with official departments and funding bodies. Island life still enforces a degree of self-sufficiency that might be equated with what used to be misnamed 'the simple life'; but happily the days when reliance on a limited range of resources implied a deficient quality of living seem to be receding. The way the island engages with the outside world is still unique.

There are no lavish museums or historical collections to structure visitors' engagement: this is the only island of such size and significance I've visited without such tourist infrastructure. Instead, the king long met every summer ferry at the harbour, doing all the work of a heritage centre and presiding over an island that continues to resist the easy route of imitation and conformity.

In his work on the postcolonial condition of Ireland, David Lloyd shows how the memory of Irish victims to the processes of modernisation can best be preserved by resisting the idea that a natural process of history created a new modern order from the ashes of a traditional, outmoded society. The danger, he warns, is that those who commemorate the victims of agricultural improvement and the mass displacement of island people do so from within the traditions of progress responsible for the destruction of lost ways of life. 'Only in remaining out of joint with the times to which the dead are lost', he insists, 'is there any prospect of a redress that would not be concomitant with the desire to lay the dead to rest.' Thoraí is a site of countless unfinished histories: a source of both cautionary tales and positive examples to help us think through the alternatives to the destructive present. Each act of resistance, though aided by the sea, could only last so long, but in a world where many seek to step back from the constant acceleration of modernity, every one is worth revisiting. The photographer of Israel and Palestine, Ariella Azoulay, writes of the 'potential histories' that persist in any place where a possible line of historical development has been violently interrupted. By this she means two things, firstly:

> the reconstruction of unrealised possibilities, practices, and dreams that motivated and directed the actions of various actors in the past. These were not fully realised but rather disrupted by the constitution of a sovereign regime that created a differential and conflictive body politic.[13]

Secondly, Azoulay's term denotes the extent to which the present is part of unfinished pasts: the violent imposition of centralised political orders can be seen not as a natural progression but as an aberration in the course of longer, more locally constituted, histories. The past is never dead but is a series of latent possibilities fracturing a present that would otherwise seem impossible to resist or undermine. In such a way the histories of places like Thoraí might – much like the histories of Ness and of machair management – make it possible to reject the present and imagine new lines of connection to the future from the past.

It was on Thoraí that my journey south collided with the northwards course of spring. A tiny vanguard of the imminent puffin invasion formed tight squadrons on the water. Fulmars and sandwich terns made the rocks as noisome as at the height of summer. Fragments of speckled blue guillemot shell showed that eggs had been laid and either predated or carelessly lost from nestless ledges. These signs multiplied as I travelled. Soon there were no geese to be seen and the familiar cackle of terns was heard regularly overhead. This was the week before St Patrick's Day – traditionally the resumption of the farming year after winter – and on the island of Gabhla tractors rumbled along the tracks with feed or fertiliser for the fields. Holes in island roofs were also being fixed: as on many islands, once-abandoned Gabhla houses are now renovated in a quiet revolution that sees sites recently left to mice and marram reclaimed for at least a little of the year. Maybe it was the influence of spring, but it was difficult not to feel some hints of optimism concerning the future of island communities, although I was soon to hear tales of incredible neglect and marginalisation.

The sea's character changed day by day, and not always in ways that seemed to match the conditions above the surface. My hips earned their keep, holding me upright through swell, breakers and tidal complexities. Talking to islanders who'd worked the sea I heard the words 'rogue wave' more than I was accustomed to, and soon saw why. After one unexpectedly brutal surf landing, where I rode a huge sudden surge to shore, I crouched to watch a handsome chough proclaim its territory beside the beach, but it took many minutes for my hands, shaking with fear, to recover enough composure to hold a camera. I was reminded how much of my winter had been spent on seas protected by the Western Isles: it was only in Ireland that I faced relentlessly unsettled waters. A dozen times or more I felt certain I'd be lost to unforgiving swell.

These violent seas, in combination with the many small islands, has created a thousand true stories of unlikely resurrections, as crews long thought drowned were discovered cowering helpless on Atlantic rocks. The sea is so storm-prone that lifeboat crews discovering stranded sailors often floated barrels of provisions their way rather than attempt a rescue. As The Times recorded in 1914, one unlucky lobster crew was twice stranded on the skerry of Roan Inish, three miles off Crohy Head. The second time, they were there for a fortnight. Fortunately, this proved to be the home of an unlikely colony of rabbits which meant the difference between survival and starvation.

Such hardships were, however, limited in comparison to the horrors these seas had seen not long before. As in the Highlands, the potato had been Ireland's miracle food. When blight arrived in Ireland in the mid-1840s, the anti-Irish prejudice inherent in British government policies turned a terrible famine into a humanitarian crisis a thousand times worse. There are tales of Irish farmers loading their tiny boats with family and a single cow to forlornly cross the Atlantic to America. But some of the most horrific reports are of much shorter journeys.

4.2 The last remains of the community of Little Bernera (off the west of Lewis). One of the most idyllic nights of the journey (little did I know that, while I was here, some visitors to Lewis, who'd seen me set off in the morning and not come back, had reported me missing to the Stornoway police).

4.3 With high swell around the Butt of Lewis, each wave hits the land before the previous one has drained from the cliff face, creating constant waterfalls down the dark rock.

4.4 On the island of Vallay off the coast of North Uist and surrounded by rich machair (in which three species of tern nest and short-eared owls quarter) stands the ruin of what was once the home of the archaeologist Erskine Beveridge.

5.1 The beautiful lone hills of Assynt. Taken at dawn from the top of Canisp.

5.2 Balnakiel (Sutherland): the heart of Rob Donn country.

4.6 Late in the evening, between the islands south of Barra, eider and seabirds had settled on the water. As I descended waves I often happened across large flocks, hidden until the last moment by the ten-foot swell.

6.1 At the end of December, storms hit. I spent Christmas in Shieldaig (Wester Ross) and Hogmanay in Achiltibuie (Coigach), sheltering from the weather.

5.3 A lone deer calf on the Sutherland coast.

6.2 Shenavall Bothy beneath Beinn Dearg (Wester Ross): a stepping stone into the mountain world of Fisherfield.

7.1 A perfect inversion, showing the cloud I'd been kayaking through for days, from the peak of Sgurr Alasdair (Skye).

8.1 Fladda Lighthouse (Argyll), surrounded by tides.

8.2 Looking down from the cliffs of Thoraí (Donegal) across to the gentle southern fields.

8.3 A thrilling and beautiful nine-mile crossing to Oileán Thoraí (Donegal); as so often, dark hugged the mainland while sunlight lit the sea. I had a sense of fellow feeling with the fisherman who was also struggling through the swell. Only when I zoomed in on the photo did I realise that he's just casually checking his phone.

9.1 In the seafoam round the Stags of Broadhaven (County Mayo), seals rush to investigate the kayaks.

9.2 Perhaps the most terrifying picture I've ever taken. Climbing up the cliff face at the Stags of Broadhaven, looking down on Llinos negotiating the increasingly rough water.

10.2 A minke whale rising to investigate the boat, a little too close for comfort.

10.3 The Dingle Peninsula from the Great Blasket (County Kerry).

10.1 Gliding past razorbill mating rituals: hundreds of birds gurgling and cooing, entirely oblivious to the kayak.

10.4 Perfect timing: the calmest of days for the crossing to the Skelligs (County Kerry).

11.1 Looking down at the meadows of Bardsey (North Wales) from the island's rough hilltop.

11.2 The Llyn Peninsula from Bardsey, across one of the most notorious straits in Britain.

On a grim winter's day in 1848 a boat known as the Derry steamer set out from Sligo along the Donegal coast towards Liverpool. The desperation to leave a ravaged coastline had been such that 190 men, women and children were packed into a steerage cabin six metres long by three wide. By the time the ship passed Thoraí, a northerly storm was thundering down the Sea of Moyle. The crew sealed the hatches and limped into the next available port. When the steerage cabin was opened, they found 'bodies to the depth of some feet . . . huddled together, blackened with suffocation, distorted by convulsion, bruised and bleeding from the desperate struggle for existence which preceded the moment when exhausted nature resigned the strife'.

As I moved on, I slowly discovered which aspects of Thoraí experience were particular and which regular along the coastlines. In the town of Dungloe I stopped for lunch at the town's supermarket, the Cope, before an afternoon in the library. I noticed with surprise that the weightiest tome on the shelves was *The Story of the Cope* (2009) by Patrick Bonar. I was still more surprised when, asking the librarian what I should read to learn of Donegal's Atlantic heritage, this was the book she referred me to.

The Thoraí system of seasonal migration to Scotland, I learned, had been general on these coastlines. Each autumn the same 'Derry boat' that had met with disaster in 1848 would take men and women aged sixteen years and older from Donegal to Glasgow for 'tattie hooking', turnip thinning and labour in mines and tunnels. But the idea that brought about the end of this unhappy human traffic was inspired by the discovery by these Irish migrant workers of Scottish co-operative societies. By 1906, a couple, Sally and Paddy Gallagher, had founded a new co-operative in the Donegal villages of the Rosses and Dungloe. Over the following decades, the Cope grew to be the largest employer in the region: it cured fish at its kippering shed, ran a knitwear factory, pooled agricultural produce through eight stores, and even secured electricity for the coastline thirty years before centralised initiatives would have provided it.

The co-operative link with Scotland soon proved crucial when the War of Independence led Londonderry authorities to blockade west Donegal, closing the only railway. Paddy (now known universally as 'Paddy the Cope') secured a trawler, filled it with eggs and set out to trade with co-operatives on the Glasgow docks who were more than willing to ignore the blockade. For seventeen more years, as the British conducted an economic war against the Free State, the Donegal railway – geared towards trade with the United Kingdom – remained compromised. Cope-commissioned steamers revived the ancient sea roads in its place. By trading along the Atlantic seaboard and circulating goods throughout the Free State they shielded the precarious economies of the west from further setbacks.

In the early 1940s the Cope, now among the most successful Atlantic co-operatives, acted on ambitious plans to revive the region's fishing industry. Three boats were acquired from Scotland, with a further two by 1953, and local skippers were trained. Creeling for lobster and catching and salting white fish were seen as the best means by which employment could be generated. 1940 was also a moment of opportunity because the able seamen of the British fishing crews that would otherwise be direct competition had been dragged from their shores, to become a uniquely valuable resource in the war effort.

Despite the loss of one boat, fishing was a flourishing aspect of the Cope's activities until the Atlantic intervened and an easterly gale in 1955 destroyed three more ships. But the legacy of the Cope boats lasted: by the 1960s, when bigger vessels arrived, there were local skippers trained to take part in the industry so that its profits weren't leached from the region. When I arrived at the Cope's main port, Killybegs, I found a site of shipping unlike any I'd seen so far. The closer I got, the more I was forced to cling to the coastline, vainly attempting to evade the unsettled washes of enormous fishing vessels. The harbour bristled with boats, arrayed in front of busy chandlers. Expanding or foreshortening their voyages according to the price of diesel, these ships ply waters off Scotland, Scandinavia, Iberia and North Africa, perpetuating the Atlantic links on which everything of this coastline's culture is constructed.

CONNACHT
(April)

N

Stags of Broadhaven

Donegal Bay

KILGALLIGAN

Erris

SLIGO

Inishkea Islands

C O U N T Y

M A Y O

Achill Island

Clew Bay

WESTPORT

Clare Island

Inishbofin

CARRIGSKEEWAUN

C o n n e m a r a

Slyne Head

ROUNDSTONE

GALWAY

Galway Bay

DÚN AONGHASA

The Burren

Aran Islands

A T L A N T I C

O C E A N

——— *Kayak route*

0   *kilometres*   25

0   *miles*   25

IRELAND IS AN iceberg. The naming of the subaqueous nine-tenths suggests a space as extraterrestrial as the otherworlds of sci-fi. The Eriador Seamount and Edoras Bank rise high beneath the waves before the black silence of the Porcupine Abyssal Plain beyond the rim of the continental shelf. 'The Real Map of Island', as marine surveyors call their image of this underwater world, charts 220 million acres under Irish sovereignty. The recording of these realms – alien but domestic – was a space-age project, begun in the year that NASA launched a venture to map Mars (1999). And the amount of new technology required to explore space and seabed is not all that different in scale.

This underwater world might seem remote, but it is linked to the realms of land and air in many ways. Birds such as petrels live lives apparently defined by air. Their acute sense of smell guides migration and directs foraging through the vaporous worlds they occupy. Yet the main ingredient of their odour map is a sulphur given off when the smallest subaqueous vegetable plankton are eaten by animal plankton. These organisms bloom at the top of seamounts, where tide-borne nutrients rise from the deep, in the same way that wind-borne moisture rises into cloud where it hits a hillside. These clouds of seamount plankton don't just attract predation but trap carbon and therefore slow climate warming.

Human commerce is also dependent, dangerously, on the deep. Lifestyles rely on the rugged few who work the rim of the ocean shelf from rig or ship. They take daily risks on a scale beyond any land work: more trawlermen-per-million die today than miners-per-million ever lost their lives while digging coal. Meals on tables a thousand miles away and fuel lighting streets in city centres began in labour on the Porcupine Bank. The biggest revelations I confronted

when travelling coasts from Mayo to Clare were the immense repercussions – political and ecological – of our economic reliance on a
world about which we know little.

Because the sea can seem so alien, and because politicians in Dublin
and Brussels have little conception of the life of seafarers, Irish coastal
communities now face crises on a scale that echoes war or famine.
Conflicts over Ireland's seas are fierce: they embody the dangers a
national, land-focused politics poses to local, oceanic ways of life. As
I visited islands off Mayo and Galway, I found myself inducted into
tensions that reveal the place (or absence) of coastlines in modern
culture.

I also saw countless illustrations of the interconnection of land
and water. Just as the refuse of the onshore world – plastics and
human effluent – spills, everywhere, into ocean, the contents of
Ireland's watery hinterland creep ashore on enormous scales. Sand,
salt and kelp blow miles inland, just as salmon and eels course
through Ireland's wet interior. After rough spells, quays and beaches
are littered with ocean death. Old ropes encrusted with goose barna-
cles – black-necked and lemon-lipped – sit beside remnants of gutted
angler fish. Their sweet fleshy cores were now on fishmongers' slabs
while discarded carapaces lay, goggle-eyed, beside my sleeping bag.
On my first night beside the Connacht sea I landed after dusk on a
long beach and settled near a log at the strand-line. Amid the rich
smells of the shore, only morning light revealed the log to be a stubby,
rough-skinned porpoise. Life and death converge on the intertidal
zones. Never have I seen so many drowned animals: several hares,
and a limp black lamb in a Mayo geo. Nor have I witnessed before
so many storm-dashed seabirds: shearwaters, guillemots and (magnificent even in decay) a gannet.

Irish poets have found this rich wreckage of the edge-zone evocative
of a heartless sea, empty of mourning and incapable of burial. Sinead
Morrissey's 'Restoration' cycle begins on a stark Mayo shoreline:

Once I saw a washed up dolphin
That stank the length of Achill Sound,
Lying on the edge of Ireland.
The Easter wind ripping it clear
Of all its history[1]

The result is a sea that looked 'wide and emptied of love'. Yet the fact that shores make us deal with death on scales to which we're unaccustomed occurs only because of the unrivalled panoply of life they sustain. The strand-line mingling of mammal, bird, fish, crustacean, mollusc, sea star, jelly, insect, plant, lichen and mould is a unique explosion of biodiversity.

Early maps of Ireland have little to say of the under-ocean world but are striking for their revelations at these shifting edges. Across decades, they reveal shorelines in natural ebb and flow, and show human interventions in the course of rivers. Since coasts and rivers were the most common form of political boundary and a place of unparalleled resources as well as the most frequent artery of travel, these maps shower them with attention. Inland spaces are often textureless, with few distinctions but those between valued and the valueless. Agricultural land, in many such maps, stretches surprisingly far up hillsides, but mountains themselves are mere lumps in profile: unshaped, uncounted and unloved. The coast, more than the land, is thronged with description, recording in words the things that resisted drawing. I was surprised how many annotations seemed inspired by wonder and sensory curiosity rather than pursuit of wealth or even safe navigation. In Donegal, I'd passed the sea cave called McSwine's Gun, described on a 1685 map as a 'place where the water howls'; in Sligo I paddled under coastal scarps 'where yearlie limbereth a Falcon esteemed the Hardiest in Ireland'; further south, 'whyt stones' are 'pointed lyke diamonds'. These annotated shorelines surely seemed like alien worlds to those in Dublin or London where such descriptions were read.

In the seventeenth century, engraving replaced drawing as the mapper's favoured mode. The result was deadening: conventional symbols for forts and harbours replaced the local quirks that sketches could preserve or emphasise. Written glosses disappeared, as did, more slowly, illustrated insets. This was also the moment when roads became focal points of Irish maps: cartography's focus was drifting landwards. Today, our most familiar landscapes in paper or pixel are grids of roads, rail or underground tunnels. Our dashboards are adorned with moving maps that take the perspective of the thoroughfare. As the geographer Robert Harbison puts it:

> On the kind of maps most people use, one feature is exaggerated at the expense of everything else, the system of roads. And yet these are seen simply as objective maps, rather than as plottings tailored to a civilisation whose relationship to the natural world is utterly and perhaps fatally mediated by cars.[2]

The first mapping projects of the modern era were symptomatic of the shift from shore to land. One of the most significant was concocted at the onset of the nineteenth century by the Dublin Society for Improving Husbandry, Manufacturing and Other Useful Arts and Sciences. The survey was organised by river basin, each of its units defined by waterways. But the society's first task was to analyse the wetlands, taking levels of large bogs that could be drained and farmed. Later, the scheme's contributors – Anglo-Irish men of means – sought routes round rivers and loughs by which railways would join the east coast to the west. Each step of this process consolidated the sense that water was barrier, not conduit; these maps both recorded and aided the drying-out of Irish life. Major mapping breakthroughs of the age of sail were now co-opted by cartographers of the land in their quest for mathematical precision, and accurate journey times, over artistic fluency and the evocation of the essence of a place. Contour lines, for instance, had been developed by surveyors who dropped weighted cords from boats to plot the relief of the ocean floor, but now mapped hills and valleys instead.

The Irish Ordnance Survey was instrumental in these changes. An arm of imperial bureaucracy, its aim was to measure the resources acquired by London in the Act of Union (1801). The imperial nature of that Union had been intensified by familiar anti-Irish prejudice: the prime minister, Pitt the Younger, envisaged a place for Irish Catholics in the new politics, but George III used royal veto to maintain a Protestant Parliament. The same prejudice compromised cartography. The chair of the committee tasked with devising the Ordnance Survey, Thomas Spring Rice, hoped to train a cadre of Irish map-makers. But Wellington considered the Irish 'too backward' for the task, so Ireland was mapped by the British army. Military surveyors – Royal Engineers, sappers and miners – became the visible face of occupation. From 1825 to 1846 Ireland was scrutinised with an intensity and precision never before devoted to a nation.

The Irish Ordnance Survey mirrored on land the Admiralty's Hydrographic Office, established in 1795. Coasts were mapped before the land because oceans were the key medium of imperial travel and areas seaward of the shore the front line of colonial reconnaissance. The Hydrographic Office surveyed British and Irish coasts while despatching gun-toting brigs and corvettes to describe the vast imperial arenas of Atlantic and Pacific. One such vessel was HMS *Beagle* from which Charles Darwin collected the specimens on which he built a career. At least to begin with, it was not that surveyors and men of science followed the flag, but that the sea roads along which empire spread had been plotted by ships of science.

Early Admiralty maps of Ireland are as full of coastal points of interest as the maps that preceded them: curiosity and colonisation worked hand in hand. Archaeological sites are noted and insets contain topographic sketches to help sailors identify and interpret points of interest on the foreshore. But as on land, the era of high imperialism saw mapping conventions formalised and the purpose of sea charts streamlined. As efficient navigation became the single goal, the coast was leached of information.

For the last century and a half, all official mapping has treated the strand-line as a barrier. In maps where the land is rich with detail (such as current Ordnance Survey sheets) the sea is a waste; in those that delineate the sea's features, the land is void. All that crosses the tideline – all interaction of communities with the sea – has become unrepresentable. The situation is both consequence and cause of our society's breach between land knowledge and sea knowledge. Our maps make us chronically sea-blind but even more drastically shore-blind.

I'd never before been aware of this breach in the ways I was on the coasts of Connacht. I was struck now by the strangeness of my own interaction with maps. Before each leg, I sat long hours with Admiralty charts, identifying sea threats and noting tidal streams in my waterproof notebooks. These notes, not the charts, accompanied me to sea. On the waves, I filtered bays, villages, hills and estuaries through Ordnance Survey land maps carried with me. Like the maps themselves, my study of land and sea was split. Each took place in a different time and space, making use of a different resource. Yet actual planning and travelling meant subverting the split resource: I read sea maps from land and land maps at sea (never the reverse) as I followed the seams where the two conjoin.

On maps, that seam is thin and appears objective: no thicker than the intertidal zone. But in practice it's wide: the subjective shore zone extends everywhere the influences of land and sea intermingle. The most prolific and profound mapper of Connacht's coasts, Tim Robinson, insists on the need 'to short-circuit the polarities of objectivity and subjectivity'.[3] By thinking of mapping as more than a quest for objective accuracy the map-maker can 'keep faith with reality', revealing the ways that wide coastal zones have been sources of wealth and poverty, tragedy and pleasure over centuries. In repopulating the subjective shoreline, Robinson shows, modern maps can begin to undermine the imperial divisions of the nineteenth century. The charts and sheets used today are still sustained by the imperial logic that laid their foundations: British servicemen had little access to local place-lore and a remit to process Irish terrain for English eyes. In this conversion, sound was prioritised over meaning: places were renamed in crude attempts at phonetic translation. This sapped land of its history. In an interview, Robinson described how the military plotted a spot they called 'Illaunanaur':

The surveyors had obviously thought that the first part of it was 'oileán', island, when in fact it should have been the Irish 'gleann', glen. But apart from making it an island when it was a glen, the rest of the name '-anaur' meant absolutely nothing in English phonetics. But in the Irish the name means 'the glen of tears' – it's exactly the biblical phrase 'this vale of tears', 'Gleann na nDeor'. And the story I heard from the local people, was that, in the days leading up to the famine when there was a lot of emigration from the islands, those emigrating would get a fishing boat to take them over to Connemara and they'd walk thirty miles . . . into Galway, where they'd wait for one of the famine ships heading for America. These ships used to sail out past the Aran Islands and very frequently had to wait in the shelter of the islands while a gale blew itself out. So they would be stationary just a few hundred yards offshore from this place, Gleann na nDeor, and people would come down to that little glen where they could wave to their loved ones but not talk to them. So the name had immense resonances and told you an immense amount about the personal griefs behind the statistics of the famine. That was very typical of what was lost in the project of anglicisation.

This contested shoreline, differently named and known by local, national and transnational interests, has been reclaimed in hosts of maps and texts that challenge official renderings of its meaning. Along the Connacht and Munster coasts there are now many 'deep mapping' projects in which communities collate the resources of place-lore to geographically reconstruct their many histories.[4] Every one of these is a project of 'counter-mapping', made in tension with official maps and aiming to rewrite the perceived meanings of the land. The availability of resources created by this process, many of which aim to erode the split between land and sea, made kayaking here an experience different from anywhere else. But the tensions that make the counter-maps necessary were everywhere evident.

South of Donegal, Ireland sweeps far to the west towards the isles and peninsulas of County Mayo. This gave me three days of shelter from prevailing sou'westerlies before the westward voyage left me as far out in Atlantic seas as the Irish or British Isles extend. I was further west than St Kilda and in waters no less exposed. This is Erris (Iar Ros, 'the Western Promontory'), labelled by Praeger 'the wildest, loneliest stretch of country to be found in Ireland'. Perhaps the land itself is wild and lonely, but the intensity of its atmosphere is conceived in a far wilder and lonelier place: on a sea where towering rollers swell and boom against torn and broken rock. This is 'one of the most treacherous parts of the western seaboard . . . a graveyard for lost vessels'.

I was joined now by my partner, Llinos, whose presence leavened the loneliness of the previous days and supplied a marked change of tone for the next week of my journey, with gregarious evenings in Irish pubs rather than cold nights on the foreshore. But each day we faced down a windswept ocean that deserved all its ill repute. As we entered open sea beyond the top edge of Mayo, thick ranks of saw-toothed swell rushed northwards and the only other life was black guillemots skipping the serrations. Cliffs fell abruptly to the sea, but the elements were still as interleaved as in an estuary. Towering peninsulas stuttered into lumps where chunks of skerry were birthed like bergs from the cliffs. These squared blocks, of vast age and thickset weight, are like nothing else in Britain. Keen tides run the channels between them, in white-cold frenzies that would be deadly should we drift off course. The melding of swell with tide, as we slid and climbed on waves, made the contoured seas frantic with movement and confusion and seemed to make the land leap and tumble. When we failed to meet the largest waves head-on, our balance and control were sorely tested. The most mind-blowing sights, however, were further offshore, where yet more lumpen rocks rose over writhing seas. Not since Mingulay and Pabbay five months earlier had I seen rock forms equal in grandeur to those where we turned from kayaking west to south: the Stags of Broadhaven are five jagged heaps, surreally precipitous and frighteningly exposed.

We reached the Stags with tides running south and swell driving north. This stirred a vortex of sea like a mile-wide whirlpool round the rocks. Although this was as bruising a day as I've ever experienced, Llinos was in her wild-eyed element as she strafed standing waves while seals rolled by. The creatures surfaced repeatedly: lithe, glistening and curious at the novelty of small boats in roaring seas (figure 9.1). Their wide eyes and flared nostrils carried a sense of urgency that made their presence anything but soothing. All was white, in a thousand shades, round paddles, boats, seals and jagged stone. Once we reached the Stags, I ploughed into the rocks, seeking respite, and scrambled up the cliff while Llinos threw herself through yet more torrents. I remember breathing far more heavily than the level of exercise merited, my chest tight with nervous energy. Hanging from the cliff edge, coursing with fear, I photographed her play amid the chaos (figure 9.2) before I lurched, anxiously, back onto water and resumed the frenzied twirl of paddles. That day my kayak felt more like the Birlinn of Clanranald than ever before: wrestling with the wind, sluiced by the waves and creaking with the effort, it ploughed into a world where lines between human, sea, boat and animal felt meaningless. Whenever the paths of our two kayaks crossed, our grins were unfeasibly wide. Glee and wonderment at the scale of the elements was amplified by the fact this was the first time on this journey I hadn't faced a rough sea alone. We didn't eat all day: not once were we secure enough to countenance taking hands from paddles. This was a magical interlude, on seas fierce enough that I might have stayed onshore if alone, and when Llinos returned to Birmingham four days later, I was entirely re-energised for the ocean trials to come.

Tales of loss and rescue abound in Erris so that archives of local news are stores of sea story. Before turning south at the Stags we'd passed long lines of sea caves that I later found in newspaper stories. In October 1997, a German retiree, Will Ernst von Below, had taken eleven-year-old Emma Murphy and her parents, Tony and Carmel, along this coast in his little currach. Heavy swell, three metres high, soon pressed them against the coastline until they were trapped in one of these caves and capsized. When the boat didn't return to harbour local fishermen and the coastguard embarked on a challenging rescue and two hours later somehow heard whistles from deep in a tideline tunnel. As dark fell, sharp gusts moved in off the ocean. The rescuers shone torches into the cave the sound had issued from and silvery patches, sown into life vests, reflected back. The local diving club were summoned to assist: Josie Barrett and Michael Heffernan dived inside, but each was disoriented in the violent sea. Barrett was recovered, exhausted, from the waves. Only when professional Garda divers were dropped in by helicopter were the Murphys reached in their subterranean huddle. But even the rescuers were now trapped in the swirling chaos of the passageway, their safety line too short to see them out. Only acts of astonishing seamanship by one of the divers, Ciaran Doyle, and two of the coastguards, Pat and Martin O'Donnell, saved the Murphys' lives. Pat made one tragic return to the cave to recover two bodies: the currach owner, Below, and the local diver, Heffernan. Awards for bravery were small recompense for a night of risk, terror and tragedy. To live on this coastline is to be accustomed to death and to be ready to improvise in the face of unique conditions, knowing fine margins and small errors separate seafarers from death.

Many villages here attest to an age when crossing wild seas was an ordinary fact of life. South of the Stags, we passed the small sea-facing settlement of Kilgalligan. Twenty miles of twisting road from the nearest town (Belmullet) the Kilgalligan road sees little passing footfall despite its staggering views across a hundred miles of coastline. Until 1942 steamers ran between Belmullet and Sligo, while currachs and sailboats ferried labourers around these smaller settlements. Making little sense as a roadside township, Kilgalligan is an artefact of Erris's risky histories: it's no surprise that nineteenth-century visitors described the local people as courageous seafarers whose natural element was winds 'which a landsman would consider a storm'.[5] They also note coast-dwellers with thick and muscular necks built over lifetimes of twisting right and left while turning small boats into the waves.

Our landfall on the day of the Stags was a tiny, isolated pier, scattered with old currachs and fishing boats. Here we met a local man walking his Labrador by the shore. He drove us two miles uphill to Connolly's pub at Carrowteige and we were soon surrounded by retired fishers and farmers, pressed into a corner of the snug by a rollicking eighteenth birthday party, complete with DJ, elaborate lighting rig, and three generations of Erris families. For an hour or so, Llinos and the Erris men swapped notes on language, discussing the survival of industries such as slate that gave resilience to Welsh, where the collapse of maritime trades had made the Irish-language cause far harder. Eventually, a weather-worn retired fisherman broke from the discussion and asked me if I knew David Thomson's classic *The People of the Sea*. I'd find, he said, that another pub nearby had been Thomson's source for folklore of Atlantic seals. He told me, as a fisherman had told Thomson nearly a century earlier, about renowned local seafarers, the Cregan family, who never drowned despite going out in the fiercest seas. A boat of Cregans once found themselves adrift after a storm and, legend has it, were only saved from shipwreck by a seal that pointed their landward route, then swam to shore and behaved so strangely that the coastguard scanned the sea and spotted a speck stranded on the skyline. Out of respect, it's said, the people of this part of Mayo became the first to stop making waistcoats and hats from sealskin. In this place, Thomson's informants told him, the worlds of land and sea became intertwined: a seal, for instance, is remembered drinking rum at a November fair, and many stories tell of evils that quickly befell any human who broke the pact with the people of the sea.

But conversation soon turned to far more recent stories of the coastline. While we were at sea, a coastguard helicopter had been tragically lost nearby, and local fishermen were engaged in the quest for wreckage, frequenting pubs to exchange information on the search: we'd arrived at a time when communities were focused on the ocean and full of concern for coastal dangers. Unlike in some parts of Scotland, though, they showed no worry about the risks we were taking: this was the first place I'd been since Shetland where small boats at sea seemed to feel as natural and familiar as the act of breathing.

We learned from our evening in Carrowteige that this is a region both ur-Irish and un-Irish: a stronghold of language and traditions, but a place where many are now fiercely ill-disposed to the Irish state. Because the land intrudes so far into ocean, there are few more contested waters than these. Oil fields and gas deposits lie close to the coast, bringing local communities and multinational corporations into unusual proximity and revealing the priorities of the state in its mediation between the two. We soon heard of locals imprisoned for protest against corporate action, and naval deployment against small coastal communities. These sound like typical tales of heavy-handed Victorian officialdom. Shockingly, they happened in the twenty-first century. Hearing them over whiskey in a packed pub I wondered whether indignation was fuelling exaggeration. Only in Galway, following up leads from Mayo fishermen, did I learn every word they'd spoken to be true and unembellished.

The context of these tensions is the selling off of Ireland's Atlantic resources. Corporations registered in Norway, Russia, Canada, the Netherlands and Spain draw greater profit from the waters west of Ireland than do Irish interests. Ireland has a quarter of the fishing grounds of EU states but just a 4 per cent stake in the fish from its own seas. Oil, gas, offshore fishing, salmon farming and even (if current lawsuits fail) the seaweed harvest are all controlled by distant interests. In a global system skewed towards scale – with huge margins for multinationals and non-existent profits for the small boat or community co-op – coastal communities can only flourish if their interests are understood and nurtured by the state. But Ireland has so long been conceptualised as an agrarian economy, in which farming interests are protected with great care, that coastal needs have been consistently bargained away.

Erris has a long history of activism against its marginalisation. In the 1950s some residents refused to pay road tax because local routes were so neglected: their taxes repaired roads in Dublin or Limerick where spending per head was many times what it was in Erris. 'They have the jet age in Shannon now,' said the neighbour of a man

imprisoned for non-payment, 'but we are still in the Stone Age here.'
Fifty years later, protests were far larger. Their arena was oceanic
in scale, because the Atlantic multinational, Royal Dutch Shell, had
arrived in Mayo seas.[6] In 1996, a substantial resource, the Corrib gas
field, had been discovered offshore from the Stags of Broadhaven.
The 1980s had seen Irish politicians such as Dick Spring work to
establish conditions for oil exploration that would benefit the Irish
people, but when few finds were forthcoming, these collapsed in
1987 into a situation that left Ireland gaining little from the vast
profits of the multinationals. These companies were able to sell Irish
gas back to the state at full market price. Dick Spring labelled these
concessions to the industry 'an act of economic treason' and, indeed,
the justice minister who oversaw the change was later jailed for
corruption. But still these unjust conditions persisted. The company
that had found the Corrib gas was Enterprise Energy who exploited
their unusual freedoms to the full: they treated Erris like their own
private property, conducting far less research and consultation than
was required for the complex task of bringing gas ashore in a populated area of outstanding environmental significance. Locals fought
back and were angered but not shocked to find council and government officials on the side of the industry.

When, in 2002, Enterprise was bought by Shell, locals rightly
feared the situation was about to escalate. The reputation of Shell
when the company came to Erris was disastrously bad. The corporation's desire to dump a defunct rig, the Brent Spar, in deep Atlantic
waters had been turned by Greenpeace into a flashpoint of eco-protest.
The rig was occupied by activists and, across Europe, Shell service
stations were boycotted. In Russia, Shell's actions were still more
contentious: the indigenous communities of Sakhalin Island had
protested Shell's despoliation of their coast, which damaged livestock
grazing and the bays they relied on for fish. The regional assembly,
Sakhalin Association of Indigenous Peoples of the North, decided

that direct action was the only response to an unaccountable multi-national that refused even to fulfil basic requirements such as conducting cultural-impact assessments. Blockades and pickets, with banners such as 'Fish are our wealth', followed. After protests directed international scrutiny towards the Sakhalin scheme, Shell's environmental approvals were revoked with over a hundred breaches of Russian law identified. Among the many shocking discoveries was that pipelines were being laid through an active seismic fault.

Shell's activities, like those of other energy corporations, were shaped not by national geographies but by the geophysics of oceans, both ancient (relict seabeds where oil and gas pipes were laid) and modern (offshore fuels and routes for vast tankers). The remarkable counterpoint to this was that their actions inspired a transnational community of protest: they consolidated connections between distant Atlantic societies, reviving an interlinked littoral that had slumbered since the age of sail. One striking instance was new intimacy between Erris and Ogoniland, a region of Nigeria 5,000 nautical miles away.

Shell's activities in Ogoniland are the most venal entries in its catalogue of horrors; events there have been described as 'the most graphic example of the "oil curse" [that linked] oil and corruption, conflict and poverty in developing countries'. Environmental destruction involved almost 3,000 oil spills as well as waste discharge that rendered vast swathes of the Niger Delta infertile. In 1989 an oil platform off the Nigerian coast exploded, adding to the deaths and despoliation. Four years later, peaceful marches of 300,000 people (almost half the Ogoni population) took place to protest Shell's plans to lay a new pipeline through Ogoniland. The Nigerian government despatched riot police to twenty-seven villages, leaving 2,000 dead and 80,000 homeless. The government then claimed that the murders of four Ogoni chiefs had been caused by the protests, so arrested nine campaigners and condemned them to death for 'incitement'. One of the 'Ogoni Nine' was Ken Saro-Wiwa, internationally renowned author and campaigner, vice chair of the international Unrepresented Nations and Peoples Organisation. Witnesses in Saro-Wiwa's trial later revealed they'd been offered jobs with Shell in exchange for false statements. Shockingly, the 'Ogoni Nine' were found guilty of these fabricated crimes and executed. The result was international outrage and Nigeria's expulsion from the Commonwealth of Nations.

All this was witnessed by an Irish nun, Sister Majella McCarron, a friend of Saro-Wiwa who had worked in Nigeria for three decades and now made certain that the campaigning cry of 'Remember Saro-Wiwa' spread throughout Ireland. McCarron's role as teacher and lecturer in Lagos and Ogoniland was the product of long-term Atlantic relations. The Nigerian Irish presence dates back two centuries, to an era when missionaries arrived by sail. Soon, Patrick was adopted as the Nigerian patron saint, 'Irish' potatoes were being grown throughout the country, and Nigerian public figures, such as the nation's first foreign minister, were being trained at Irish universities. By the 1960s, in what has been called a 'religious empire', Irish personnel occupied positions in every stratum of the Nigerian church and every division of its provision, running 2,419 primary schools and forty-seven hospitals. Today, there are Nigerian expats in Atlantic cities such as Cork and Limerick and Irish communities in coastal cities such as Lagos. The arts of the two nations have intertwined. Yeats echoes through Nigerian literature, from dedications in poetry such as that of Christopher Okigbo, to the title of *Things Fall Apart* by Chinua Achebe. But the reverse influence is perhaps more significant. Issues raised by Nigerian writers, such as Achebe's question 'What must a people do to appease an embittered history?', inflect Irish efforts to understand the colonial legacy. Irish authors wrote regularly to Achebe, sometimes expressing their wish that someone might illuminate the famine or the troubles with the power with which Achebe expressed the condition of colonial and postcolonial Nigeria.

This all meant that when Shell arrived in Erris many locals saw the company's name as synonymous with the plight of Ogoniland. They'd read dozens of national newspaper articles and donated money in church fundraising campaigns. Sister Majella had founded Ogoni Solidarity Ireland who began, in 1996, an annual Saro-Wiwa Seminar. In the year that Shell bought the Corrib gas field, this event saw Komene Famaa from Ogoniland draw parallels between the recent past of his people and the immediate future of Erris. Soon, the Ken Saro-Wiwa Archive was established at Maynooth University. As press reports put it, 'there is a little corner of Ireland that is forever Ogoniland'.

Over subsequent years, Irish activists undertook many forms of peaceful protest including changing the street names round Shell's Dublin offices to Ken Saro-Wiwa Street. A Dublin MEP, Mary Lou McDonald, congratulated these campaigners: despite the prominence of Saro-Wiwa's name,

> still the Irish government continues to collude with Shell in the destruction of the sensitive environment of the north-west coast of Mayo, while doing deals which effectively hand over the country's oil and gas resources to Shell and other multinational energy companies. To give away such valuable natural resources in a time of economic uncertainty is inexplicable. Ireland should be harnessing its own natural energy reserves not handing it over to multinationals virtually free of charge.

In Mayo, locals set out to scrutinise Shell's activities in the ways a responsible government, less dazzled by multinational wealth, might have done. They drew attention to the high-pressure pipelines that would be built three times closer to residential homes than safety guidance suggested. They pointed out the weakness of peat bogs through which such pipes would be laid. Indeed, in 2004 a graveyard and road were lost from the pipes' proposed route as peat and turf slid downhill. Some locals even began to name themselves 'Bogoni'.

From the start of the project, residents had witnessed developers begin clearance and construction before permissions were granted. Local houses and land were given to the oil companies under an unprecedented scheme of compulsory purchase, but locals refused to watch their homes destroyed until they'd seen permissions in full and confirmed their legality. Five residents of the Erris village of Rossport were jailed for non-compliance and 'the Rossport Five' became local heroes in life as the Ogoni Nine had in death: crowds of protestors guarded their land through all the time they were locked away. Activists set up Rossport solidarity camps, while Shell multiplied its security and the state committed 3,000 Gardai (a hundred times the usual local compliment) to protecting Shell's interests. At the same time, locals collated the many dimensions of potential impact on the coastline. New attention was paid to the animal species using the bay and to the potential impact of seismic interference. Analysis was devoted to the place of fishing in the local economy and to the damage to fish stocks that would be caused by release of waste products. Local contacts with Ogoniland and Sakhalin Island allowed conclusions to be based on evidence of Shell's past actions.

The historic coastline played a contested role in these campaigns. Pipelines would rip through beaches, field systems and famine-era structures that were mapped in greater detail by local tradition than by official cartography. One flashpoint came in 2002 when, while still waiting for permissions, developers began to excavate the beach where proposed pipelines would make landfall. An eighty-four-year-old local, Mary Philbin, brought an injunction against them because their digging destroyed a cillín (a burial ground for unbaptised children). No cillín was marked on Ordnance Survey maps, so developers and Dublin authorities could wield the official record of the Irish landscape when they insisted this stretch of coast had no cultural significance. Such claims escalated the tension, because they showed that it was not just a single site that was 'violated by the most arrogant and uncaring forces', as Philbin put it, but the whole edifice of local tradition. Maps made by the occupying force of the British army were being used to undermine a place, associated by locals with great psychological trauma, whose very existence had been predicated on secrecy. The question that was now at stake was how, if at all, coastal oral traditions could constitute evidence in a court system built on the recorded facts and written words of historic imperialism.

This question would never be fully answered: no satisfactory solution was found to the implicit dismissal of local knowledge contained within the legal system. But these debates did destabilise Shell's approach. Shell tried constantly to return the debate to specialised scientific language in which locals could have little stake; they even labelled the protests 'a sustained assault on scientific reality'. Yet the complex array of voices that confronted them meant Shell were unable to dictate terms and in the words of one local, their 'aura of power [was] undermined'. It was in the year I travelled here that Shell's incursions were rebuffed and the corporation finally gave up.

Every mile of coast south of Erris is evocative of a famed aspect of Ireland. Alone again, I soon rounded a row of vast cliffs at the largest of the islands, Achill, once the hub of the basking-shark hunts that provided oils which, before the era of Shell, powered Ireland. Here, days after I passed, a storm restored a beach that had been missing for thirty years: tons of sand appeared overnight on bare rock, as much like a fairy tale as the sudden theft of the beach in the 1980s. From Achill I moved into the stretch of coast versified by one of Ireland's greatest nature poets. Just as Norman MacCaig spent summers in Assynt and winters in Edinburgh, Michael Longley escapes the cities for warm, wet weeks among the Mayo waterbirds. Like MacCaig, he praises moor and ragged coastline, locating transient tracks of dunlin and sanderling and peering through the heather to find delicate delights like the lesser twayblade. The distances evoked by the fauna near Longley's chosen village, Carrigskeewaun, add to its legend: it is 'a townland whooper swans / From the tundra remember, and the Saharan / Wheatear'.[7] He marks the shifting boundaries of this tide-digested place whose gestures to eternity – forts and burial mounds – are easily 'erased by wind and sea'. Longley dreams of being buried high on the relict shoreline: entombed in a plateau that had once, when 'the old stone-age sea' came far inland, been an Atlantic isle.

Leaving the land facing Corrib didn't mean escaping the tensions between conflicting systems of power and knowledge. Indeed, crossing from Mayo into the most historic coastal corner of County Galway – Connemara – meant moving into a region famed for its tensions of old ways with new. This is a place now mapped with greater inventiveness and attentiveness than anywhere else in the Irish and British archipelago; yet all its recent mappings have been countercultural acts of subversion aimed at the blind spots and falsehoods of the official record. This was where the experience of kayaking was made unique by the cartographic projects that record lives lived across the strand-line, rejecting the familiar breach between land and water and showing how, in the words of Tim Robinson, 'the dense record of life is scribbled in the margin of the sea'.

The Aran Islands, the Burren and Connemara form a single unit of past seagoing life gathered round Galway Bay and once crossed by hundreds of currachs and Galway hooker sailboats. Geologically spectacular – mountainous and complex – the region is known as 'the ABC of earth wonders': Aran, Burren, Connemara. This zone has a marginal history, its population swelled by people expelled from sites of power through Ireland's stormy past. It's the place Cromwell had in mind when forcing Irish-speaking Catholics to 'hell or Connacht'. The language is still at its strongest here, with a depth of oral custom that extends to every otter haunt and twinge of tide. These are the heartlands of the sea god Manannàn Mac Lir, whose mythic burial place was a dry Connemara glen until Lough Corrib sprang from his grave to flood the valley. A host of bays and islands here are named for him. Oileán Mana was created when he required a landfall to save his mortal daughter from a storm. The very name, Connemara, is derived from a people who considered him their ancestor: in English these were the 'Conmaicne of the sea', in Irish the Conmhaícne Mara.

Connemara is the peat-cloaked northern edge of the earth wonders: a winding granite coast of extravagant fractal length. It is gnarled, dark, and barred from the inland east by the high twelve bens, but life explodes into leaf and flower at the fringes of the many rivers that pour through a thousand grooves from the mountains. The shoreline itself is swathed with an astonishing diversity of seaweeds, including rare blanket beds of pepper dulse (míobhán): the most delicious of them all. The rich, intoxicating flavour of this 'truffle of the sea' (like garlic, game and mustard all in one) is bound tightly to my memory of kayaking there.

Across Galway Bay from Connemara, Aran and the Burren are seen as thick karsts of pitted limestone that gleam grey white. They are staggeringly rich in flora and fishing but their treeless, peatless shores yield no fuel. They are the strangest places to paddle because they look so different from the sea than from onshore. If you're stood in its midst, the Burren is a world of exquisite delicacy. Leaf and flower hide in every rocky crevice: to shoegaze here is to meet a miracle at every step. One morning, I lazed on the limestone scarp and watched spring gentians unfurl as the dawn met their deep blue crowns. Then I plunged into the deeper blue of the Atlantic and a world less rich than raw. From the sea, the Burren is not delicate but armoured: steely scales like hard reptilian hide clothe everything; the flora in the crevices is not just invisible but unimaginable.

Life in this amphibious ABC always depended on moves around its regions. Fuel, food, song and story were constantly exchanged because every corner lacked what the others had in plenty. And because of the glacial stripping of the region, these circulatory histories are legible underfoot: geologies and fossils, holy wells and forts ensure that every scrap of shore speaks of centuries. As I paddled, unexpected castle ruins appeared, mirage-like, behind knolls. From the water, burial mounds stood out that would otherwise have hidden among dunes.

Tim Robinson is both the crux of Connemara's modern mapping and the scholar to interpret most compellingly the richness of *dinn-seanchas* ('place-lore'). Trained as a mathematician, he was a successful London artist before the epiphany that 'art is the opposite of money' sent him scrabbling to the islands of the Irish Atlantic.[8] What was planned as a fleeting escape became an existential recentring. Robinson has become as much a philosopher of place-lore as a scientist or artist, his achievement less in record-keeping than meaning-making. When he and his wife Mairead arrived on their Aran sojourn, they found themselves 'hijacked' by Atlantic coastlines and histories. Both learned Irish fast and, over decades, deeply, until most great honours of Irish language and culture have been bestowed on them.

Within weeks of arriving, however, the question of how to represent a place so misconstrued by surveyors began to gnaw. Robinson's art became a science of reintegration and reconciliation: a cartography built not on measuring with new precision but on exploring the complex reciprocities between humans and their world. He wandered the angular stone walls of Aran through rain, shine and sea mist absorbed in the question of how a single segment of earth's surface could be known in depth. 'I wore the network of tender little fields and bleak rocky shores into my skin', he told the ecocritic Pippa Marland, 'until I could have printed off a map of them by rolling on a sheet of paper. But I was always aware of the infinity of ways in which the place exceeded my knowledge of it.'[9]

The maps that resulted are rooted in earth science: built on geologies the Ordnance Survey ignored. They are layered with place-lore: re-Irishing the shore with a millennium of accumulated custom. And, because the cartographer 'must be faithful to more than the measurable' they are inked in freehand, reviving the artistic subjectivity that imperial cartographers sought to banish from the craft. They say as much about time as space. The past is here on every scale from the birth of land to the recollections of the living, each a crucial component of a holistic quest. There is never an overarching narrative or an easy structure that imposes meanings. Indeed, Robinson is as critical of the Christian logic of the ninth century, which forced narrative structure onto the swirling well of old Irish lore, as he is of the rigid formulae of imperial cartography. His sites are thus opened to diverse interpretations rather than having meaning pinned through them. There is something determinedly anti-Enlightenment in this approach: his researches don't cohere into bright-lit histories of linear cause and effect but are 'points of attachment of the historical web from which one can grope back along the strands into the darkness'.[10] This Enlightenment principle he resists is the effort to make science universal. The sciences, in that vision, elucidated laws operating everywhere the same and have dismissed every shred of evidence that points to what the historian of exploration, Dorinda Outram, calls 'the embeddedness of the natural order in a particular land'.[11] This was the same problem that scarred the Outer Hebrides: the belief in booming cities that the principles that had brought them success could be exported unchanged to ensure success everywhere.

Robinson also restores the pre-Victorian counterpoint of map and prose. Every sheet has a gazeteer unpicking the implications of each name. But short gazeteers were just the start. The scale of *dinnsean-chas*, and the sense of responsibility to the past that Aran inspired, meant Robinson's cartographic works became 'preliminary storings and sortings of material for another art, the world-hungry art of words'. Profound and poetic books soon dug beneath the topsoil to 'find chinks in Time' and endow the maps with unrivalled historic depth. The books – two for Aran, three for Connemara and many more of thematic essays and stories – are without rival as the richest body of writing on place that the north-east Atlantic archipelago has ever inspired.

Robinson's practice is grounded in two unfashionable scientific tools: walking and talking. These are, he insists, the best methods to ensure 'the moment by moment reintegration of body and world' on which comprehension of place must be founded.[12] He has written at length on the intellectual potential of these tools, but a statement he made to Marland about the mundane benefits of travelling on foot is revealing of how both his practices operate. He echoes Trevelyan's appearance at the crofter's door, but in a refraction rich with empathy and social conscience instead of aristocratic ego:

> Once, a wealthy friend with a big car offered to help me in my explorations of Connemara. Since I wanted to revisit a few remote glens I accepted, and we roared off. Then, 'I must call in at that cottage,' I said, and we squealed to a stop. I knocked at the door, but apart from a twitching curtain there was no response – whereas if I had sweated up the hill, fallen off my old bike at the gate, asked for a bucket of water to mend a puncture, etc., all the lore of the valley

would have been forthcoming over tea in the kitchen. But even bicycling is inferior to walking in this context. To appear out of the thickets behind an Aran cottage, or scramble down from the bare moon-mountains of the Burren into a farmyard, is, I find, a disarming approach, introducing me as obviously unofficial and dying for a cup of tea.

I'd mused for some time on the openness I'd met with from those who'd be justified in scepticism, even hostility, at an English visitor's efforts to describe their communities. Robinson's short spiel was the first thing I read to indicate how strongly my waterborne bedragglement might have spoken in my favour.

One morning, Robinson's role in this scenario was reversed: a lone kayaker squelched to his door. I arrived at 11 a.m. at the quay in Roundstone village, where black guillemots bobbed under the window of the Robinson home and studio. I hoped we might talk for an hour, but the hospitality was such that it was two meals and nine hours later by the time I left. We looked at maps, prints and dolphin skulls before lunch. Then we drove to a beautiful cottage shaded by elm and oak. Tuaim Beola stands beside a still black pool where tides push Atlantic salt into clear floods falling from the Connemara hills. It was the home of friends of the Robinsons and once the dwelling of the philosopher John Moriarty, who passed away exactly a decade earlier. Today's event was a small remembrance in poetry, song and story for a writer whose ideas on the soul and sensuousness, deity and beauty, inflect many Irish approaches to human and non-human nature. Here I met for the first time the poet Moya Cannon, described by Moriarty as 'the finest poet in Ireland', whose verse had inspired my engagement with these coastlines from the start, and whose kindness would give me introductions to ecologists, artists and musicians in Galway city.

Moriarty and Robinson were the closest friends and starkest opposites. Each loved the Connemara coastline and each wrote prose of ecstasy and awe in its praise. For Moriarty the landscape was an act of God and a cure for the ailing soul that was best illumined with the language of spiritual transcendence. He took his friend to holy places where Robinson played along with the required responses. But Robinson's world view is built on determination never to read the immediacy of real terrain in terms of an ineffable beyond. His sense of wonder is coterminous with the material world; but his praise of place is no less visionary or exultant for that fact.

This is an attitude I've seen more in writing on Ireland's Atlantic than anywhere else. When Seamus Heaney gazed on the sea roads of Irish saints he 'saw only the secular power of the Atlantic thundering'. Glorying in planetary force and the sensory impact of seas that could seem timeless, he struck a pose of intense and sublime material poetics. If Heaney was the psalmist of that precise but heightened mode, then Robinson is its prophet and high priest.

This attitude is key to understanding a body of work driven by devotion to the ritual and litany of these small stretches of wildly diverse coastline. But the attitude is rarely directly stated. Marland's interview is brief, but more revealing on this theme than countless other expositions of Robinson's craft. She asks whether it's important for humans to recognise their material nature and Robinson answers with dazzling conviction. Just as he 'short-circuits' binaries between subjective and objective mapping, he renders any distinction between the religious and secular impossible to draw:

The notion of a disembodied afterlife strikes me as blasphemous. If mentality can be sustained without matter, then the vast fantastic fabric of material existence, from the atomic particle to the neuron to the galactic cluster, which it has been my deepest delight to learn something of, is a waste of space and time.

During the day of Moriarty's memorial, we discussed Robinson's rapturous fixation on the coastline. It isn't just distinctions between religion and secularism, mapper and mapped, art and science, or English and Irish that his work re-fuses, but also that between sea and land. In every genre he seeks brackish edges like the pool by Moriarty's cottage 'where land and sea entwine their twisted fingers'. It is this rejection of the shore as boundary that meant his work transformed my kayaking. These are the only modern maps I know where coasts are centres not just of space but of the map's emotive world: they are clad in thick cloaks of annotation that the land is not. Walking the cliffs of Aran, where the Robinsons made their first Irish home, provided his epiphany: 'a cliff face is ignored by the conventional map', he noted, since it appears as a line of infinitesimal breadth.[13] But a real cliff 'was a wide province of the islanders' mental landscape, a theatre of anecdote, tradition, boast and dream'. Much of life is vertical, invisible to surveyors' eyes, in amphibious communities where differential planes of land and sea conjoin. In drawing his Aran sheets Robinson indulged a subtle stretching of the shoreline, laying the cliff face almost in relief. A heady plunge becomes a space five millimetres wide, thick enough to swathe in place-lore. This is one example of how precision is misleading and inaccuracy most closely represents the truth.

It would be some time before I reached Aran, but when I did Robinson's map made the bare precipice as scrutable as a city. The clifftop fort of Dún Aonghasa, 'the most magnificent barbaric monument in Europe', is riddled with pre-Christian lore. Its building is ascribed to Aonghas, leader of the wild Fir Bolgs in the time of myth. Walls six metres high were constructed on the cliff edge. Now some parapets lie in ocean depths while the rest stand on the 'beetling brow' above. They tower 100 metres over any kayak that nears the Dun's open and undefended seaward edge. No hint can be seen from here of the shiny signage and gravel paths on the landward side and the view feels spectacularly Bronze Age. Robinson finds the Dun equivocal: the Aran site most compromised by tourist infrastructure and retaining its power only from certain angles.

> If the setting sun is riding into the bay on the backs of the waves, illuminating the vastness of the opposite precipice in golden detail, while the solemn recession of promontories beyond goes back step by step into rose-petal impalpability on the western horizon, then the setting is definitive: Dún Aonghasa, heavy with centuries, dreams upon a pinnacle of another world.[14]

The other world is Celtic myth, which he reads as an elemental faith. The Dun's stormy grandeur shows primal elements at war: light meets dark and airy heights meet ocean depths. The ancients transmuted earth, sea and sky into a 'cosmogonic politics': the battles and matings of gods and goddesses. Robinson's material rapture is prefigured in their world view.

But the Dun itself has no more presence in island lore than the cliffs beneath it. It stands over An Sunda Caoch ('the Blind Sound'). This inlet leads to nothing and haunts the dreams of islanders as a place of shipwreck and the point where the isle will one day split in two. The O'Flaherty family lived across the Blind Sound from the Dun. Their home was the place the island went to dance, sing and tell tales. Two sons of this house are now renowned for turning Aran's cliffs into novels and short stories. Liam's dark romance, *The Black Soul* (1924), begins in the first storms of a hard winter when sour winds make for green and bilious seas and cause a stretch of Aran cliffs to fall. But it was his brother, Tom, who turned his pen most often to the crags. He published two collections of short stories, *Aranmen All* (1934) and *Cliffmen of the West* (1935), which record the place-lore of his home.[15] Tom calls the eastern cliffs beneath Dún Aonghasa Aill an Eala ('the Cliff of the Swan'), while the western overhangs are Carraig an Smail ('the Cliff of Perdition'). The latter name was coined when Aran boatmen found a body in their nets and, recoiling at its smell and appearance, returned it to the water; it was the corpse of a neighbour, fallen when fishing wrasse from the clifftop. The boatmen's failure to bring the body home for burial rankled for decades.

A small cave below the Dun is Poll an Tobac ('the Hole of the Tobacco') where goods from wrecked ships were hidden from customs men. The *ailleadoir* ('cliffmen') who inspired Tom's writing worked these rocks each day and knew routes along *aragaint* ('ledges') and *strapai* ('cliff-path') that led to marine, mineral, avian and salvaged wealth. These were vertiginous pathways, movement subject to a skewed, sideways physics, yet they were as familiar as the island's fields or lanes and just as productive. Robinson stresses the geography of resources in cliff-face place names such as Leic an tSalainn ('the Flagstone of Salt') indicating spots where life's necessities might be found or made. But the routes are now disused and lost in rockfalls, and Robinson, following Tom O'Flaherty, is rare in recording this tradition:

> I see these heroes as bent, wheezy little old men with a comic turn of phrase, for that is how the islanders I talk to recall them; it seems that those vigils on the windy ledges were conducive to wit as well as to catarrh.

Some, he suggests brought the savagery of the cliff face home: one man raised peregrine and raven chicks to force into cockfights. Never, without reading Robinson, would I have recognised these overhangs and ledges as a human theatre full of character and domestic drama. How many other barren cliffs I'd passed beneath, which had no Robinson nor O'Flahertys, had in fact been equally busy sites of production?

It was from speaking to islanders, however, that I learned the extent to which the fishing community here is threatened with the oblivion of the cliff men. Like the oil and gas of Mayo, the marine resources of Galway have been lost thanks to the state's landward vision. The major setback for Aran fishing was the accord struck in 1973 when Ireland joined the European Economic Community. European institutions aimed to establish the waters of member states as a shared resource, granting nations with short coastlines a stake in seas elsewhere. This would be an admirable goal were it not for a debilitating flaw: coastal communities depended entirely on freedom to make flexible use of local resources. Irish islands, and small Irish ports and harbours, have, for over four decades, been faced with conditions that threaten to end their existence as anything other than hives of holiday homes. European rules lacked the subtlety to prevent huge vessels registered in distant ports from emptying vast swathes of the Irish Atlantic. These are resource-raids inspired by short-term profit in contrast to the long-term custodianship that comes with communities of small-scale fishers. And enforcement wasn't keen enough to prevent the dumping of billions of tons of dead herring as factory ships played fast and loose with the stipulations of their quotas. Much of the problem wasn't the fault of European administration, but a result of Irish negotiators' total lack of any sense of the needs of Atlantic islanders.

The last few years have seen powerful films transform public awareness of these issues. First there was *A Turning Tide in the Life of Man* (2014), which recounts the efforts of an islander to reason with bureaucracies. Then there was *Atlantic* (2016), which compared the threatened communities of Ireland with the lost industries of Atlantic Newfoundland and the thriving ocean industries of Norway. And major research projects now aim to breach the divide between these communities' understanding of the sea and the very different vision of distant institutions. These include the Irish Fishers'

Knowledge Project at the National University of Ireland, Galway. But it remains to be seen if anything can be done that will prevent the need for a future writer in Robinson's mould to recover a sense of the lost Aran life of today.

The most revealing set piece in all Robinson's work demonstrates the need for any understanding of community, of place, or of the relations of production, to be rooted in time. It evokes inspiration from the Aran strand-line rather than cliff edge. Wading into ocean to watch dolphins as they wove through sea, Robinson mused on relations of human movement to time and space. The present and thousands of past processes exist in every step, he realised, and the task of decoding a single metre is beyond the wherewithal of humans:

> I waded out until I was within a few yards of them. Their beautifully easy plunging motion entranced me – the way they moved through the waves was itself wave-like; it seemed to express the perfection of their adaptation to their habitat. I began to wonder what the human equivalent would be. Could one imagine taking a step, just one step, even on the world we inhabit, that would be as adequate to the ground it covered as is the dolphin's wave-like plunge to the wave it traverses?

Pursuing this logic of attentive being, which recalls no other attitude so much as that of a medieval monk, Robinson's work offers an orientation of extraordinary precision but also inspires an experience of extreme disorientation as the moment swings into deep time and back.

I'd thought most of the temporal contrasts Robinson draws on the day that summed up my wet and windy time in Connemara. This was days before I passed beneath the Aran cliffs, and I still wasn't sure what implications Robinson's writing might have for my journey. I was rounding a unique headland that is neither cliff nor shore but as complex as either. This was one of very few places that Robinson, by necessity rather than inclination, surveyed by boat instead of on foot. Slyne Head is the westernmost extremity of County Galway. Its squat dark rock is broken by rampaging seas into scores of scattered, salt-sprayed islands. The sky as I paddled hung low and intense. Black clouds were heavy with vapour, ridged like rough water, and goaded by the same sou'westerly that whipped the sea. Exposure like that here is rare except beneath tall cliffs. In swell of any size this low land is lost: it drops beneath the paddler's view as if sunk in sea. Ragged sheets of wave loomed, coiled, and fell across the kayak, solid, forceful and relentless. The effect was claustrophobic: like contracting walls of nightmare cells. The horizon was near enough to touch and could never go unwatched for a moment.

Sometimes, however, there'd be a fleeting glimpse above the scarps of sea. In those moments close perspective transformed to cinematic sweep. Horizons bright with golden light spilled between pewter sky and iron ocean. There were sea-strafed reefs where towers of white foam grasped like stalagmites towards cloud as solid as the ceiling of a cave. The dark sea bore ranks of marching furrows, flecked with white, a thousand strong. I tried in vain to blink sea spray from tired eyes whenever my tiny world was transformed into this vision of stupendous forces. I felt smaller than I've ever felt before and my little paddles spun like the wings of a moth fluttering moonward. I thought of Robinson and the words with which he evokes each step in a Connemara field as a point of access to vast inhuman scales of space and time. His phrase – 'the immensities in which each little place is wrapped' – captured the paradoxes of close focus in the second biggest ocean on earth. Robinson's eyes and limbs are always straining towards dimensions inconceivably larger and unimaginably smaller than the human: it was here, as my senses dealt with droplets and the ocean dealt with me, that it struck me how awe-inspiring Robinson's philosophy of place could be.

I landed at Slyne Head as the clouds collapsed into downpour and cleansed three days' salt-rime from my skin and clothes. The views slammed shut. Although low light warmed the rain for a little longer, grey melted into grey and ground receded into shadow. The island was cryptic with granite that slowly quit glittering. Two lighthouses, less than half a mile apart, mark this out as a site of danger. One is a grim black light at the island's west that feels like something from Tolkien, the other a stunted grey tower at the east. An even clearer symbol of exposure is a strange long corridor like a fragment of labyrinth: a walled pathway that bisects the islet so that walks across it can be made in storm-force winds (though it might be fifty years since the trudge was last made in earnest). There are unexpected steps and platforms amid the rabbit-dug land and brackish pools. Bright creatures perch beside rain-pelted puddles; gradually, they reveal themselves through the drifting grey veil to be buoys blown ashore by gales.

The night felt like winter. I was wrapped in wet darkness until the huge black light whipped by. Its beams were unwelcome three times over: they intensified the dark between each sweep, they lit constellations of falling water that I hoped in vain to forget, and they marked time through a deluge that felt interminable. The surprise was that there was something comforting about this spot, precisely because it was humanised by Robinson's record of shoreline life. It couldn't be wild when populated with his warm cast of fishers and farmers for whom each skerry was familiar. Their mocking, laconic responses to fearsome weather made it impossible for me to feel sorry for myself. His is a vision of these islands that's close to us in time yet radically different from the ways of life in Dublin or Belfast or Cork. Never idealised but always enchanted, never mystical but full of mystery, the ordinary and domestic are revealed to be astonishing and wild, and the wild and astonishing, domestic.

I cruised a few days later straight into the heart of Galway, having rounded the great promontory of Ireland's mid-west and dealt with the most challenging seas of the whole journey. Galway was like a fireside in winter or a lighthouse in a storm: an oasis of arts and sociability on the edge of the Atlantic. A mercantile town of great antiquity, Galway long provided Norman invaders with protection from both the ocean and the Irish but is today a haven of Irish-language culture. Maybe my feelings were shaped by the wild context of the preceding days and by the introductions Moya Cannon had granted me, but I've never been so instantly enamoured of a city on first visit. The presence of urban centres such as Galway and Limerick in deep clefts of the land mass distinguishes the Irish Atlantic from the Scottish and explains, in part, why Irish culture once faced the ocean more resolutely than any other part of the British and Irish Isles. It may also help explain the greater resilience of the Irish than the Scottish language after 1800. My next few days were more intensively social than any others on the journey: breakfasts, lunches, coffees, dinners and drinks ran one into the other, each with artists, writers or scholars from the National University of Ireland. The wrench of moving on was palatable only because an approaching spell of calm promised to ease my passage through the finest skerry-scapes in Europe.

MUNSTER
(May)

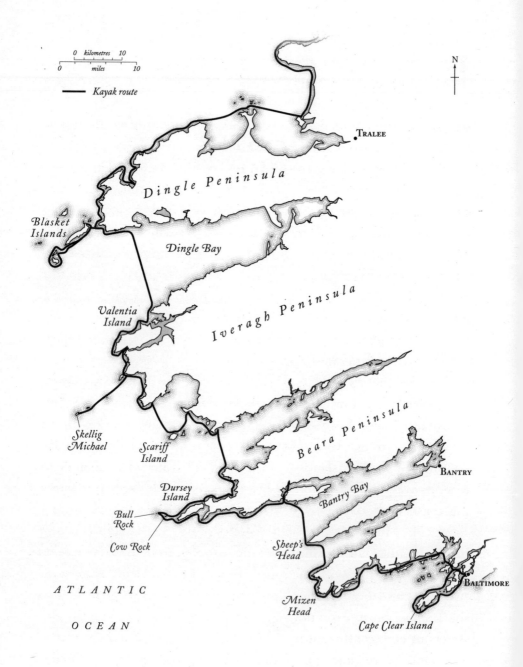

0 *kilometres* 10

0 *miles* 10

—— *Kayak route*

•TRALEE

Dingle Peninsula

*Blasket
Islands*

Dingle Bay

Iveragh Peninsula

*Valentia
Island*

*Skellig
Michael*

*Scariff
Island*

Beara Peninsula

•BANTRY

Bantry Bay

*Dursey
Island*

*Bull
Rock*

Cow Rock

*Sheep's
Head*

•BALTIMORE

ATLANTIC

OCEAN

*Mizen
Head*

Cape Clear Island

I ARRIVED IN Munster at the perfect time of year. As spring waxed, shearwaters arrived. These large, piebald birds, with long cylindrical bills of gun-metal grey, transformed the experience of the sea. By day they glided noiselessly a foot or two above the waves, drawing uplift from the sea's contours, appearing silently at my shoulder. At dusk they sat together in vast rafts, conjoined into ocean obstacles, and quiet unless spooked into clamour. After dark they began to vocalise: island nights were haunted both by ghostly wails from their burrows and the zip of their long wings cutting the air beside my head. Sometimes I'd round a headland to find myself in choirs of cooing razorbills. I'd drift, silent and ignored, through the rituals with which they remake pair bonds every spring: tapping bills, nuzzling, gurgling and swimming to and fro with feathered faces pressed together (figure 10.1). I'd never before witnessed these extraordinary displays of affection.

Days lengthened and seas warmed till the high shadows of gannets were matched below the boat by dark forms of pelagic fish and mammals. I soon heard rumours: gangs of basking sharks, fifteen or twenty strong, were lumbering north along the Kerry coast. Among the Blasket Islands (Na Blascaodaí) I watched an alien creature, its waggling nose golden-bright in the last light, flap towards the coast. Only days later did I discover it was an infant basking shark. The Munster names of this species relate to its habit of lounging in fine weather with its fin above the water close to shore: *ainmihide na seolta* ('monster with the sails'), *liop an dá* ('unwieldy beast with two fins') and *liabán gréine* ('great fish of the sun'). Oil from this shark's liver burns exceptionally clean and smokeless, so that the city streets of eighteenth-century Galway glowed with shark: 'sun fish' is a name twice deserved. The creature's plenitude was such that thousands were killed to light the roads: in 1774 a whaler named Thomas Nesbitt harpooned forty-two from a single small boat.

Far out among offshore rocks I paused in mingled fear and wonder as round-faced Risso's dolphins burst through waves and overleapt my boat. This was the first time I'd seen this species. Up to four metres in length, their presence was far more imposing than I'd ever imagined. They were ballistic in velocity and purpose: dark torpedoes of pure energy that ripped through still seas. The Risso's bull that came to investigate at close waters was young and almost black, lacking the pale scars that mark older individuals out as alphas. Whales were more familiar company and were the Risso's antithesis. Their small fins topped the gentle arcs of their huge bodies in the roaring tides near headlands; they were oases of slow calm amid waters in fast-forward. On my second day off Kerry I was caught in the tiny shining eye of a minke whale that showed its pleats as it slowed almost to stillness and observed the boat; I managed a single photo of a minke fin (figure 10.2). Whenever bubbles rose in the lea of land, tiny, glistening porpoises were sure to rise behind them. It wasn't now unusual to see three species of cetacean in a day. On the afternoon I met the Risso's there were five. And there's no experience I know that compares with being locked in the gaze of creatures that seem to live on planes of existence so remote from ours.

It's no surprise that Irish culture has for centuries been full of these ocean megafauna. The sea itself was a place of transcendence that looms large in old Irish writing. Poetic inspiration and spiritual insight were thought to occur at the edge of water. Next to heaven in the list of God's great creations was 'the white waved sea on earth'. The bright and shining ocean was a mirror of the heavens; a halfway stage between ordinary earth and transcendent air. In words attributed to St Fionan it was '*in muir múaid mílach*' ('the noble sea, teeming with life'). In lines attributed to St Columba, the 'waves above the shining sea' sing to God 'in everlasting sequence' so that anyone aspiring to sanctity must gaze 'on the manifold face of the waters'. But even this saint, often so focused on his deity, wished to be by the Irish Atlantic for the same experience sought by countless modern visitors: 'that I might see its mighty whales, the greatest wonder'.

In the older Irish tradition, waves were the blowing manes of the family of Manannàn Mac Lir. These sea gods' dishevelled tresses were loosed, and turned white, when the gods were wild with joy or anger. But this spiritual sea of Celtic lore was also run through with the imagery of cetaceans. A long seafaring poem attributed to Rumann mac Colmáin, who died in 747, defines the ocean by its beasts with just the surface spume being god-locks: when storms arrive

> The pallor of the swan covers
> the plain of whales and its inhabitants;
> the hair of Manannàn's wife blows loose.

Even before Heaney translated *Beowulf*, the medieval kenning of the sea as 'whale-road' swam his work, as in the ten exquisite Glanmore Sonnets (1979) which conjure the blown horns of trawlers returning to harbour for shelter from a storm: 'Sirens of the tundra / Of eel-road, seal-road, keel-road, whale-road raise / their wind-compounded keen'.

Several Irish myths, such as that of St Brendan, recognise that whales are not so much animals as floating ecosystems. They feature cetaceans as land masses on which a party of saints or heroes can dwell for days before discovering their isle is animate. Others feature whales as foes, threatening boats and swallowing heroes whole. In the Irish Cinderella tale, Trembling, the hero, is pushed from a sea cliff by her sister, Fair, who then sets out to seduce the handsome prince while her virtuous sister pines in a whale's belly. Trembling is swallowed and regurgitated three times before she's found. Only the Irish telling of this tale, otherwise common to many cultures, turns on a whale's appetite. Irish Jonahs abound. When Marianne Moore, American daughter of an Ulster mother, wished to assert her identity as a woman of Irish heritage she wrote an essay entitled 'Sojourn in the Whale' (1915). The offending whale stands for England, English, empire and patriarchy. However, the poem turns on the phrase 'water in motion is far from level'. All will not run smoothly, Moore insists, for the Anglo-male whale: a renaissance of cultures like Irish will result in the regurgitation of the things it has consumed. In the surreally understated terms of the critic Fiona Green, Moore's whale suffers from the 'digestive difficulties that trouble the colonial project'.[1]

But Irish culture also features whales as symbols of beneficent strength. Ireland has no official 'national animal' like the Welsh dragon or English lion but the whale has often been an unofficial sigil. In the early days of the modern Olympics, when Ireland was politically British, many athletes would not represent the occupying power. The most famous were a group of weight-throwing policemen who, for the USA and Canada, won twenty-three medals between 1896 and 1920. These transatlantic strongmen took the natural name 'the Irish Whales' and became a global phenomenon.

Many shoreline sites are named for cetaceans, such as Cuasín na Muice Mara ('Little Inlet of the Sea Pig') on Clear Island (County Cork), or Ros a Mhíl ('Promontory of the Whale') from which ferries depart for the Aran Islands. These evoke the presence of real rather than symbolic whales, often connected to historic strandings. But the practical meaning of these rich masses of meat has changed dramatically since the sites were named. On Inis Mór a whale stranding would once have meant largesse in flesh, bone and blubber, but by the early twentieth century the sight of a vast carcass on the shore inspired disgust not joy. A song, which seems only to be recorded in Tim Robinson's archive in Galway, recalls the day when:

A big reward was offered for to bury him, for we
nor any other men could haul the monster out to sea.
You'd want stout Coll McMorna, Oscar and Conan Maol,
and even they could hardly stand the stench of rotting whale.

A crowd came o'er from Galway, determined to be rich
'We'll cut the body up', they said, 'without the slightest hitch',
but something must have scared them, when once they saw the
   whale
for home they went as if they had the devil on their tail.

Some came from Connemara, and brave enough to stand.
'We'll set alight to him', said they, 'and burn him on the strand',
but they were unsuccessful and one was heard to say,
'You'd need the Ardmore witch to come and spirit him away.'

In an era that rejects whale commerce these leviathans have become symbolic of environmentalism and blue ecology so that the hulk of a whale speaks on a new scale: not of local windfall but of global tragedy. It is the awe-inspiring antithesis of urban industry and the icon of oceanic otherness.

The fact we think of 'the whale' at all is a symptom of this shift. The ethereal song of humpbacks is conflated with the scale of the blue whale, the intelligence of the sperm whale with the friendliness of the grey to create an idealised chimera that is a gentle totem of the ocean. Always, it seems, 'the whale' is baleen grazer not toothed hunter. Indeed, I once experimented on my students, asking 'What do whales eat?' All answered simply 'Plankton.' Not one showed awareness of the range of whales with their many diets; none seemed willing to recall that toothed whales existed. Like the Irish west itself, 'the whale' is guardian of old values otherwise lost: survivor of a truly wild ethic and intelligence. It is to nature as Fenimore

Cooper's last Mohican was to culture: the noble savage on a frontier in mid-collapse before forces of homogeneity and profit. In modern cultures the whale is the ocean aboriginal who lived through millennia of prehuman peace. It is a survivor from the age of mega-fauna whose onshore analogues are not so much elephants as sauropods. Whales are, in this vision, watery Adams and Eves who never left their blue Eden and are unstained by the sin with which myth explains human nature. These fictions of the whale are no less significant and meaningful for being false.

Whales are thus living histories that embody a particular attitude to time. This attitude elevates everything ancient. It casts grand perspective on the fleeting triumphs and disasters of the human present which are made to look like surface froth in contrast to what the whale sees: the slowly moving deep. Though there were once many whale-hunting stations along the Irish shore, these have almost been erased from memory and are mostly beyond identification. In north Donegal I'd found a briar-bound shoreline structure, believed by a few locals to have been a whaling station, but now only ever listed as a ruined salt works. The fact that whaling was a matter of local subsistence in the recent past has been partially erased from consciousness, the hunt presented instead as a pursuit of Greenland, Norway and the high Arctic.

It seemed as I travelled south from the Blasket Islands like every cetacean in Ireland had been pooled in my path. Reaching those islands felt like passing through a gateway between cetaceanless and cetacean-filled worlds, so it was no surprise to find their stories on the shore. The Blaskets themselves are a famous site of porpoise-eating. This small cetacean's Irish name, *muc mhara* ('sea pig', a name also occasionally used for the smallest and most common Irish whale, the minke), conjures the same culinary associations as the English (derived from *porcus* and *piscis*: pig-fish). Their meat was often simply known as pork. Porpoise bones are found in Neolithic kitchen middens, and references to their slaughter appear in texts from the ninth to nineteenth centuries. Nowhere were they more prized than on the Blasket Islands.

The Blaskets are the first of the south's offshore splendours, followed by the Skelligs, the Bull and the Cow rocks, Cape Clear Island and the incomparable Fastnet Rock. These imposing island groups are the tips of huge peninsulas: long, fluttering strands of land that are the very raggedest fragments of the frayed Atlantic edge. The Blaskets are high and airy. Some are mound-like, others bristling with jagged rocks (figure 10.3). The views along the Dingle Peninsula from the top of the westernmost isles (Inis na Bró and Inis Mhic Aoibhleáin) are among the finest vistas that human eyes can access. To the east is a narrow wedge of shapely mountains that slides into the distance with a perspective deeper than any sea. Wrens, stonechats and rock pipits animate the foreground, flitting between the wild castles of huge rocks and through the unmortared remnants of island homes. Close, in the Atlantic, are the westernmost fragments of Ireland, the awesome rock pyramid of the mini-skellig, An Tiaracht, and the boat-shredding barriers of the Foze Rocks. The only person to have lived amid these immensities in recent years was the wheeler-dealer politician Charles Haughey who bought and retreated to Inis Mhic Aoibhleáin from the financial scandals that erupted periodically around him.

The easternmost island, the Great Blasket, is a tragic site, where hearth, infield and moor stand empty, only different from the day the islanders left in a few respects. The animals that roam the slopes around the village are now just sheep, where cows, pigs, hens and donkeys (the latter often referred to as 'the voice of Blasket') wandered as freely through island living rooms as across the land at large. The houses' roofs of thatch or tarred canvas are mostly now blown away and the richly decorated chimney breasts collapsed. When in daily use, these houses were fragments of sea on shore:

their floors were made from driftwood and boats' hatches, sometimes still barnacled, and sprinkled each morning with white sand from the strand below. Visitors record having mackerel nets spread over driftwood as makeshift mattresses. Now that the houses are empty, these floors have been reclaimed by the turf and grass of the land, their sandy floors blown back to the shoreline. Where the living village had redistributed land, shore and sea – disturbing their boundaries – the island's abandonment had quickly restored the order in which each element existed in its own domain.

The last residents, who sailed away in the 1950s, wrote a host of accounts of their island lives, full of the psychological trauma of decline and displacement. Among these, Tomás Ó Criomhthain's *The Islandman* (1929) is most famous. Ó Criomhthain records his suspicion that small boats mysteriously lost near the islands were often overturned by whales and sharks. He also describes a day in 1890, when he and his elder sister wandered the shore. They were surprised and frightened to see a huge pod of porpoises careering towards them. Soon, three island currachs arrived to push the creatures ashore, but six mainland boats came from the village of Dunquin to shepherd the pod from the islands and catch them for themselves. The Blasket boats fended off their rivals, beaching the rich prey and striking the killing blows. Overjoyed islanders ('in those days you could hardly get anybody to exchange a porpoise for a pig') were soon as bloody as the porpoises themselves, struggling with the task of lugging unwieldy corpses to the village for salting.

> My father's face was red with his own blood and the blood of the porpoises . . . I jeered at the old woman whenever she came along with a creel full of porpoise meat balanced on her rump. You might imagine that she came out of one of the porpoises herself with her creel, she was so thickly smeared with blood. But she had earned a meed of praise, for she nearly killed the captain of one of the Dunquin

boats with the blow of a shovel. The islanders had no lack of pork for a year and a day . . . There was no risk of me forgetting that day even if I should live to be a hundred. Everybody you saw was crimson with blood instead of being pale or swarthy.

This account evokes the serendipity and flexibility of island life. No one present, it seems, had seen a pod like this before, the sea pigs usually imagined to be solitary. But the islanders acted swiftly when they saw a resource to ease their winter. In a single day, the island menu was transformed for months to come, thick steaks of porpoise added to the sizzling mix of pollock, wrasse and bacon roasted over turf. The unforeseen bounty of the sea reduced the pressure on subsistence for a year.

As I left the Great Blasket behind, porpoises were plentiful, having long outlasted their human hunters. A small motorised rib approached me and the couple onboard told me of a large group of basking sharks five kilometres west. But I had to continue south. Each day's kayaking was now a long leap between the stepping stones of Kerry and Cork peninsulas. But each crossing was teeming with life. The first took me through legions of gannets and auks to Valentia Island. This is a place forever associated with surveys of ocean fauna thanks to the pioneering naturalist Maude Delap (1866–1953), whose life has recently been given fine historical treatment by the geographer Nessa Cronin.[2] Over a lifetime of ocean observation Delap elucidated the life cycles of jellyfish, the distributions of sea plants and the movements of cetaceans. She and her sister Constance rowed their small boat beneath Kerry cliffs recording sea temperatures and collecting specimens of plankton, anemones, shellfish and seaweed. Descriptions of the home Maude shared with her widowed mother and three sisters depict in particular detail the sea-flavoured fug that filled it: her room, like her mind, became an intense distillation of the Atlantic. She preserved molluscs and fish in pungent formalin; she kept her jellyfish in jars of brine; bones and seaweed dried on sills and tables; and the lilies she sold to fund her research didn't hide but intensified the atmosphere with ammonia and spice.

Lacking formal education or high social status and living at a distance from the learned societies of Dublin or London, Delap's rise to recognised expertise reveals the unique ways in which far-flung Atlantic outposts became centres of scientific fieldwork. Their striking feature is that the wealthy men of elite urban institutions were always more reliant on unpaid and self-taught fieldworkers than vice versa. The usual elitism of the scientific establishment collapsed in the face of geographical difference. Delap was given memberships such as to the Linnean Society; she wrote at least fifteen articles, and was showered with praise by the metropolitan scientists who collated her work and that of her peers. She was offered at least one paid, high-status post in England but never took positions that entailed leaving both her family and the island that was her laboratory.

Delap's remarkable commitment to chronicling the shore, however, soon led to her appointment as first official recorder of cetacean strandings for the Irish south-west, operating under the aegis of the British Museum. The museum began efforts to map strandings in 1913 and was soon producing theories on migration and distribution of whales as well as on the sensory world they occupied. Receivers of Wreck (the officials who administered law in relation to maritime salvage) were given forms on which to record descriptions and measurements of every carcass seen. But self-taught naturalists who lived in regions far from cities were the best resource: the people who really knew the shores. The museum equipped these researchers with the latest cetacean manuals and bespoke supplements written in a spidery, almost illegible, hand. Delap joined their ranks in 1920 and soon made the most remarkable find of any operative. This was the first complete body of a True's beaked whale (the only previous, incomplete, finding was from North Carolina in 1913).

The aftermath of the discovery is revealing of many of the era's scientific dynamics. Delap made the identification, and had the carcass (weighing close to two tons) moved to her garden. She sent the head and flippers to the museum for verification before burying the body in her asparagus patch. Two years later, having finally agreed that this find was exceptional, the museum requested the remainder. Delap dug up the year's veg and despatched the complete skeleton by sea. After several weeks she received another missive requesting the missing pelvic bones (at the time these were thought to be vestigial in all whales, but are now known to play roles in the mating practices of some cetacean species). The asparagus was excavated once more before a telegram arrived admitting an error: the True's beaked whale had no pelvic bones. As in many other disciplines, boundaries between professionals and amateurs, 'authorities' and 'assistants', did not imply distinctions in knowledge, expertise or effort. Marine biology was still an improvised affair with much groping in the dark, whoever conducted it. Committed figures like Delap were subordinate only in financial resources, not expertise, to the chairs and committees of venerable institutions. Delap's access to the ocean was a resource worth more than any institutional endowment.

In 2003, fifty years after her death, Maude Delap returned to Valentia. Her notes and drawings were laid out on tables and her jellyfish floated through the church where her father had been priest. Visitors strolled beneath an enormous image of her forehead as they stepped across the boundary of her mind. The occasion was a film by the artist Dorothy Cross and her marine-biologist brother. Cross has interrogated the creatures of these seas as closely as anyone, building art from whalebones and fusing the nature and culture of the shore by stretching the skin of a stranded basking shark over ribs of a wooden currach. In *Medusae*, the film shown on Valentia,

Cross aimed to see the jellyfish through Delap's devoted eyes. They became mythic and meaningful when subjected to sympathetic gaze. Delap was also made mythic, not through a narration of her life but in the effort to use her eyes and ears as the viewers' windows on the world.

This was a fitting tribute not just to Maude herself but to those, often also women without independent wealth, whose lifelong work in branches of archaeology, marine biology and every other field of coastal knowledge did more to elucidate these shores than the histories record. The work of Sydney Mary Thompson on the glacial geology of Kerry, of Matilda Knowles on marine lichens, or Mary Ross on the geologic links across the Sea of Moyle, is only now beginning to be recovered. As scholars such as Nessa Cronin are showing, the Irish Atlantic provides an exceptional range of material for the ongoing project that the American geographer Mona Domosh sought to encourage when she made her 1990s call for a feminist historiography of geography. Domosh showed that 'women's ways of knowing', formed by the social conditions of femininity in Victorian Ireland, run through the construction of geographical and ecological knowledge despite the tendency of traditional histories to record only the input of professional, institutional, men. It isn't, Domosh insists, just stories of heroic women that are lost in traditional narratives, but a whole way of conceptualising nature and science, land and sea, which was instrumental in the formation of today's outlooks. Watching jellyfish through Delap's eyes, until every figment of their diverse and delicate anatomies comes into focus, conjures a sense of the attitudes and approaches that the professionalised sciences once endeavoured to exclude from their origin stories.

The south-west of Munster has been, through history, the most consistently outward-looking region of Britain or Ireland. Its identity was built in the gold and silver ages of the Mediterranean world and its fates later shaped by the transnational Catholic alliances that were formed in the heyday of Iberian power. Irish origin tales begin in Spain, Palestine or Egypt: Irishness was founded on tales of Galician magicians, Phoenician traders and biblical patriarchs. But Atlantic influence, generated by the monks of the west, soon flowed back into the wellsprings of the faith itself. And there is no doubt what the greatest physical monument to the early Irish integration into Mediterranean culture is. The stupendous skerry of Skellig Michael is a place of almost infinite romance and grandeur to which more fine words have been devoted than to many regions 10,000 times the size.

This was my next oceanic stepping stone after Valentia Island. I approached over a glassy sea on a day of glittering calm, passing hosts of seabirds and their chicks. The surreally pristine conditions (my only risk May sunstroke) left me laughing to myself like an ancient anchorite who'd spent too long in sun-kissed isolation in the Alexandrine desert. I had the perfect weather to imagine this place continuous with the olive groves and papyrus fields where biblical rivers meet the sparkling Mediterranean. Only as I reached the rocks did a little protective cloud gather, permitting me to gaze upwards at their stupendous columns (figure 10.4). There are two huge sker-ries here: grey-green Skellig Michael is the most celebrated of all the early-Christian sea rocks, while its prickly, white and silver twin is a vast gannet colony. That 'skellig' is derived from the Irish term for 'splinter of stone' seems fitting, though these are splinters beneath the fingernails of a vast deity, not anything corresponding to human scale.

This was a place where my kayak didn't, as it did elsewhere, separate me from other modern travellers: the skelligs unite all visitors through similarity of experience. Because visiting boats are so small, and no landing can be attempted by large vessels, all pilgrims sit small and low in the water as if in supplication at these immense altars in the ocean. They all struggle ashore from little boats after silently drifting beneath huge wheels of gannets with nothing more than the ritual spinning of paddles or regular splutter and thrum of a tiny engine. Not just a kayaker, but every visitor, is thus at least a little attuned to the early-Christian holymen.

There's something essential about the modesty of all these small-boat approaches to the Skelligs. These rocky eruptions from the water are unique in bombast: vast and arrogant as sea-born Babels. Yet no human has ever attempted to build grandly on them. Even the lighthouse on Skellig Michael is a squat affair, almost embarrassed at its intrusion: at twelve metres tall it is a fraction the height of other lights on dramatic rocks, such as Fastnet (fifty-four metres) and Skerryvore (forty-eight metres). The older buildings aren't like sailing ships or cruise liners but are kayak- or currach-small, built in rather than atop their element. The three flights of stone steps weaving up to Skellig Michael's twin pinnacles look, until close, too small for human use. The tiny monastery sits 600 feet above the water, both diminished by distance and dwarfed by spires of rock behind it. Skellig Michael's man-made elements thus feel strangely consonant with Little Skellig's gannetry – a tiny human hamlet perched as precariously, and knitted into the cliff as completely, as the vast bird city next door.

The monks who once populated this rock were the great Atlanticists of their age. The monastery was founded in the era when St Fionan studied with saints David and Gildas in Wales before sending his own pupils, Brendan and Columba, to build new seats of learning throughout the seaboard. It was in the monastery's heyday that the Irish monk Dicuil composed his world geography in Iceland, drawing on the work of his fellow Irishman, Fidelis, who had sailed to the

Holy Land and back. The Irish were, for centuries, renowned as the most learned people in Europe. The era of the scholar saints' expansion throughout the east Atlantic was the era commonly labelled the 'Dark Ages'. They are often referred to as 'keeping alive' the delicate 'flame of learning' after the fall of Rome. But this was not just a little interlude between Rome and Renaissance, to be so condescendingly dismissed. It was a movement of intellectual life away from cities and into oceanic institutions: an embrace of the Atlantic as a site of learning, as well as a flight from cities whose fortunes were fading in a world no longer trapped in the iron grip of an immense imperial power.

From its earliest recorded moments, Irish Christianity featured an asceticism comparable to that other great wellspring of monkish devotion, the Alexandrine desert. Indeed, legends (which there is no good reason to dismiss out of hand) claim that Irish monks had visited the desert fathers, and that seven Coptic Christians were buried in Ireland after making the return journey. Irish children earmarked for a religious life were stripped of everything familiar. Denial of family, hearth and homestead became a prime mark of piety. This was apt psychological preparation for the bleak conception of intellectual and religious vocation that living in a twelve-strong community of men on Skellig Michael entailed. Scraping away at the scant earth to grow vegetables and scouring the cliffs for seabirds and their eggs, scholar-monks pursued meaning in a place sacred to the God of storms. There would have been little or no cooking or heat here, because there were no fuels besides fish oil. Their religion was famously more attuned to older traditions of nature worship than Christianity elsewhere, maintaining a cycle of feast days – Samhain, Imbolc, Beltain, Lughnasadh – that simply transmuted old gods into new saints and allowed for the presence of God in everything. The remarkable continuity this entailed has been used to explain the fact that no one at all was martyred in the Christianisation of Ireland. Seamus Heaney characterised this unique Irish theology as the ability to see the god in the tree as well as in the creator Father and redeemer Son. But in the early golden age of monastery and hermitage, perhaps the god in the sea was a more pertinent point of reference than that in the tree. The leading historian of Skellig Michael, Geoffrey Moorhouse, refers to one particular verse as summing up the continuity of Irish tradition.[3] Said to have been penned by an ancient Irish poet of the old gods, Amargin, it was preserved through use by Christian monks and translated by the first president of Ireland, Douglas Hyde. It begins:

> I am the wind which breathes upon the sea
> I am the wave of the ocean
> I am the murmur of the billows

And it ends:

> Who is it who throws light into the meeting on the mountain?
> Who announces the age of the moon (if not I)?
> Who reaches the place where the sun couches (if not I)?

Even the writing with which monkish ideas have been transferred to us was done with products of the ocean: fish oil and the whites of seabird eggs bound colours to the page, while the colours themselves included purples extracted from shellfish. The labour of salt-worn hands thus runs through every illuminated manuscript now venerated in the national museums of Dublin, Edinburgh and London. But there is, of course, no need to choose between sea and tree as the favoured dwelling of this deity: when St Patrick was called to play his celebrated part in the Christianisation of Ireland, it was both wood and water that spoke to him in his dream: 'I heard the voice of them who live beside the wood of Foclut, which is nigh unto the Western Sea.'

It is one of my chief regrets of the journey that I didn't spend the night on Skellig Michael, communing with its uniquely oceanic variation on the Christian deity. But I had no up-to-date forecast and dared not risk the weather turning overnight while I sat miles out in seas famed for brutality. I climbed to the monastery and lingered there as long as I dared, amazed – even after all I'd seen – that these fragments of cut stone looking out on nothing but ocean once stood as a central pillar of western culture. Then, when I could wait no more, the rich yellow sun almost touching the flinty horizon, I skipped down hundreds of narrow steps and launched onto a glistening sea. It was completely dark by the time I reached the mainland and improvised a spot to sleep on a tiny beach beneath an old stone quay.

Every day that followed brought stupendous rocks, such as the Bull and the Cow, as splintery as miniature skelligs in seas that seemed as full of cetaceans as the whale-roads the monks had known. I was now among sea rocks that had been the last strongholds of the Gaelic chieftains against Cromwell. Every habitable islet is scarred by tales of slaughter: 300, mainly women and children, were executed on the Dursey Island cliffs, while one holy man, Diarmaid O'Sullivan, who had attempted to equip his countrymen with armaments he shipped from Spain, was run through on beautiful Scariff Island where I spent a calm night under the brightest starlit sky of the summer. Indeed, the weather held till, four days after the Skelligs, storms delayed my final crossing and I found myself a little room to rent in Baltimore before a final wild kayak to Cape Clear Island in the extreme south-west. I had somehow survived the Irish Atlantic. Despite the phenomenal calm that held while I kayaked Cork and Kerry, the thing the Irish coast had impressed on me the most was its ferocity. By any sensible calculation, this uniquely brutal sea would seem the least suited I've seen to travel by small boat. Yet seafaring knowledge has been preserved here through millennia by the small communities on island and peninsula so that tiny-boat traditions, fragile and threatened as they are, have survived on a scale that only Shetland can even, quietly, echo. There is nothing inevitable, the Irish case shows, about the ways in which our societies have turned their backs upon the ocean.

BARDSEY TO THE BRISTOL CHANNEL
(June)

N

NANT GWYTHERN

TREMADOG

Llyn Peninsula

PWLLHELI

HARLECH

ABERDARON

Ynys Enlli/
Bardsey Island

I R I S H   S E A

ABERYSTWYTH

C a r d i g a n
B a y

NEW QUAY

CARDIGAN

FISHGUARD

ST DAVID'S

Skomer

Grassholm

Skokholm

PEMBROKE   TENBY
Caldey Island

——— *Kayak route*

0   kilometres   20

0   miles   20

THIS MONTH WAS the least Atlantic of my journey, an interlude undertaken because no approach to the seaboard could
be complete without Wales and Welsh. Everywhere I kayaked this
month was guarded by Ireland from at least some vectors of the
ocean's momentum. But on the coastlines south from the long Llyn
Peninsula, where Dublin is the nearest capital and Cardiff many
hours' drive away, the coast's relationship with oceanic elements is
not straightforward. These shores are protected from all winds but
the sou'westerly gales that rush through St George's Channel. Then
they partake in full Atlantic fury. This mixed essence isn't just
geophysical, but cultural, historical and paradoxical. These regions
feel far closer to historic land empires and centralised cultures, from
ancient Rome to modern London, than anywhere I'd travelled before.
Yet they maintain their linguistic and cultural differences more
strongly.

The first site of such mixed exposures and histories is Ynys Enlli,
the island of Bardsey, and I began my month's journey with a wind-
wracked week on its shores. On the coasts of Wales, less ragged
than those I'd paddled so far, large islands are rare. Because their
existence is more remarkable, the last century has seen poets, novelists, artists and naturalists converge on their shores. These writers
– from Ronald Lockley to Brenda Chamberlain – were instrumental
in creating the modern idea of the island and defining the condition
of being sea-surrounded. Today's 'islomania' thus has a distinct
Anglo-Welsh flavour, made by English sojourners on a handful of
rocky havens.

My month would begin and end in those small outposts. But
between Bardsey and the Pembrokeshire islands of Grassholm,
Skomer, Skokholm and Caldey lies the long, low and islandless sweep

of Bae Ceredigion (Cardigan Bay). This is an entirely different kind of space, a huge buffer zone between fragments of archipelago. It is a space peppered with places once famed for shipbuilding. What are now small seaside towns then saw more ship construction than Cardiff, writing towns like Pwllheli and Barmouth into the histories of empire and slavery.[1]

The experience of kayaking here was also different. I came to think of this buffer zone as a transition between two very different parts of my journey. Roaming Bardsey felt like being on Scottish or Irish islands: intensely welcoming of those who made it out across the water, a place where my kayak acted as a ticket of admittance to an island community and I could talk afternoons or evenings away with people fascinated by the sea in all its aspects. This was the last place on the journey that I experienced such a thing. I was met on islands further south not by people interested in what I was doing but by wardens attempting to charge me landing fees or send me away unrested: people who showed little understanding of the sea or its traditions. I'd reached places where the sheer number of visitors has eroded tolerance of, or interest in, their stories. This was a gulf between communities and visitors that cannot be crossed as readily as further north or west. Being here in tourist season meant I had no choice but to experience the South Wales coasts and their communities as tourists experience them.

I was also now more often wandering concrete than marram grass, because the human presence on these coastlines is more brazen than that on any straightforwardly Atlantic shore. Seas are tame enough for conurbations to take root. These towns haven't always embraced the water: early houses were much like those further north, facing inland with backs resolutely upon the ocean. The sea here was seen as a practical place – even a kind of infinitely enlarged cesspit – not an aesthetic site or leisure zone. Only at the end of the eighteenth century did massive new demand for seaside resorts turn houses round and the historical trajectory of these towns begin to align more with those of inland England than of the frayed Atlantic edge. Town after town now squares up to the incoming elements, gazing out across open water instead of cowering in inlets or estuaries like Atlantic conurbations. Their long strands are visible from miles away. The roar of jet skis rends the sleepy bays, and plastic buckets and other lost accoutrements of the holiday camp bob by at sea.

Yet the water here is still an immense force in winter and the coast shifts and crumbles before its power. Pwllheli spends millions in redistribution of sands which the sea moves, regularly restoring the town's beach after its seasonal drift along the coast. Aberystwyth cracks and creaks before winter storms and then rebuilds its shattered seafront for the summer. This ambiguous coastline is thus full of compromise and contradiction.

These mixed meanings stretch to the human culture of the coastlines. Names of coastal towns are a dual litany of Welsh cultural heartlands and anglicised holiday destinations. The population of the Llyn Peninsula springs from 30,000 to 150,000 for a six-week season as surfers gather at Abersoch and caravan parks fill with guests. But among the 30,000 are the people who make these towns the core of the second-largest language in the islands: the cultural success story of which so many Gaels are envious. Each town strikes a different balance between the competing demands of this fractious duality, and it is astonishing how different places a few miles apart can feel for this reason. The nature of their names hints at the geographies of culture. The towns of the north retain their single Welsh names: Aberdaron, Abersoch, Llanbedrog, Pwllheli, Criccieth. But for every mile travelled south the chance increases that an anglicised name will rival it: Abermaw is also Barmouth, Cei Newydd is New Quay and Ceredigion Cardigan. My opportunities to practise a few words of Welsh would reduce with every landfall until, in Pembrokeshire, far more French, German and Dutch than Welsh is heard in coffee shops and pubs. Names such as Penfro (Pembroke) are heard more rarely than their anglicised forms. The places in the southern reaches of this month's journey have Latin names as well as Celtic and Saxon. I began to see remains of the great historic culture that, of all those in the west, was most uncomfortable at sea. Rome once built a great edifice, Moridunum Demetarum ('the Sea Fort of the Demetae people'), on this gold-rich coast. There are rumours of Roman forts still further west, such as Menapia on St David's Peninsula, though their archaeology has not yet emerged.

The fact the Romans braved these coasts more readily than the Atlantic has shaped the ways scholars approach them. The Irish Sea is, in the words of the leading scholar of the post-Roman aftermath, Peter Brown, 'the Celtic Mediterranean'; Brendan Smith, in his survey of the later medieval archipelago, uses the same analogy, adding that this 'tapestry of interdependence' around the Irish Sea formed a 'true economic and cultural region'.[2] The Romans encouraged this cross-fertilisation by settling Irish peoples onto the South Welsh coast and in so doing created the kingdom of Dyfed that would dominate South Welsh history for five centuries. Strange artefacts resulted, such as stones carved, uniquely, with both the ancient Irish ogham script and Latin. This easy circulation has never ended. In the nineteenth century, Bardsey fishermen rowed their catch long distances around the Irish Sea. They set out in overcoats stiffened with tar, to combat both the elements and fatigue, and when they arrived in Liverpool or Wexford were known as 'Welsh penguins' because of the strange gaits the tarred coats gave them.

Important though Irish Sea histories are for this region, modern language and culture dictate a geography that is distinctively Atlantic. As the Romans once bound South Wales into a vast land empire, the Welsh language binds these coasts into oceanic geographies. When I explained my journey in Pwllheli I was told of busloads of Ness children arriving for cultural exchanges. I recalled that several Irish-speakers I'd met had spent weeks learning Welsh at Nant Gwythern, the language school perched on the Llyn sea cliffs. In July I'd learn that the major institutions of Cornish-language revival have been much like franchises of existing Welsh entities.

This solidarity between languages and communities is wide in scope: Wales is bound linguistically and logistically to Galician and Breton coasts as much as to Cornwall and Ireland. In the 1990s, the European Commission established the idea of an 'Atlantic Arc' in which Wales, as the largest site of Celtic language activity, was intended to be pivotal. This was one of eight 'Euroregions' conceptualised to encourage transnational responses to shared cultural needs. Small regional initiatives and grand transnational agendas thus spotlit the significance of these cultural enclaves where geographic schemes formed around 'the nation' thrust them to the margins: language and marine resources were the twin issues at stake and Welsh and Breton interest groups were the most active in championing the initiative. Here was my month's greatest paradox: that a place so pivotal to any vision of what north-east Atlantic culture might mean was not actually Atlantic. And the month began on the most paradoxical of islands.

It is often said that the Isle of Arran is 'Scotland in miniature', echoing every facet of the nation's mountain zones in one rocky eruption from Argyll seas. Ynys Enlli (Bardsey, 'the Island of the Tides') has a similar encyclopaedic quality, melding all association of 'islandness' in a single small package. Towering cliffs are stitched to bouldery shores and beaches. These switch suddenly into dense wetland, lush valley, rich pasture and a unique rough upland whose floral extravaganza is punctured by the burrows of thousands of seabirds (figure 11.1). The seams between these scenes are abrupt and obvious. Arguing that Bardsey was too complex to ever comprehend as a whole, the photographer Peter Hope Jones catalogued the island as an array of partial images reflected in water: bright, intertidal rockpools reflect serrated coastal stone; ephemeral rainwater ponds mirror ferns, gorse and heather; deep, clear wells reflect the massive open sky. The patchwork colours and textures are echoed more spectacularly still at sea. Small zones of smooth or ruffled water, like irregular fields of pasture, form a tapestry of wild tides with different rhythms and tangents. Through every day the sea is a slow kaleidoscope, each zone shifting from flat sheen or twisting frenzy to flat frenzy or twisting sheen. These wild tides are like gates that open and close seaways: a seemingly safe doorway to the island can be slammed shut in an instant.

Although emblematic of the conditions of islandness, Enlli is also unique, its modern history haunted by the received meanings of an illustrious but overbearing past. This was once among the most sacred places in the British and Irish islands, with a presence in medieval verse second only to Iona: *'neud uchel wendon gwyndir Enlli'* ('white waves make loud the holy land of Bardsey'). The tides were a revolving door for pilgrims whose bones now enrich the earth: everyone who has spent much time here has stories of unearthing them. The names of coves and headlands, such as Porth Meudwy ('the Hermit's Harbour') express holy histories as eloquently as the medieval abbey and Victorian chapel at the island's centre. The inscription on a Celtic cross, built to commemorate the most hands-on modern landlord, the third Lord Newborough, makes certain that visitors can't miss this historic identity: 'Respect the 20,000 saints buried on or near this spot'. Newborough even bestowed upon one of the island men the office of king, inventing historic institutions where he couldn't reinvigorate existing ones.

Despite Victorian emigration to America, the island population survived into the twentieth century. Residents sixty to ninety in number were often supplemented by the crews of large passing ships. Enlli was famed throughout the region for a rich diet based on pigs, hens and the crops of its fertile, bone-filled fields. In the 1870s the small island houses were demolished at Lord Newborough's command and replaced with pretty whitewashed dwellings. He was so enthusiastic concerning the life lessons the island offered that he sought to build a model community. This became a kind of sea-bound Bournville or Welwyn Garden City: a beacon of good living in the ocean.

This initiative drew Enlli to general attention, giving the island an exceptional celebrity. Visiting journalists occasionally filled their narratives with the progressive scenes Newborough hoped for, but more often with anecdotes of rustic eccentricity (habits as uncouth as the staggering faux pas of serving shellfish warm with potatoes instead of cold with bread). But there were deeper meanings to the newspapers' responses to the new Enlli. In an era concerned with the apparent diminishing of religion in both public life and literary culture, it was as the 'Island of Saints' that Enlli gripped imaginations. Through the institution of the parish, religion had for centuries been the primary lens through which British landscapes had been organised and understood. The secularisation that occurred in the nineteenth century was a matter of bureaucracy, not of changing numbers of believers (the vast majority continued to profess faith whether privately or through public acts of commitment). The significance of the parish as a division of cultural space had been chipped away at consistently from the 1830s onwards as governments sought to centralise administrative functions and bring all within the secular reach of London. The intensely sacred space of Enlli thus became a symbol of what was being eroded, and the starkest contrast to the newly functional conception of land.

It helped, perhaps, that the people of this era often interpreted faith in terms of oceans, and saw its crises as storms, receding tides, or a general drying-out. The most famous poem of the mid-Victorian 'crisis of faith', Matthew Arnold's 'Dover Beach' (1867), sees its speaker gaze out to sea:

> Listen! you hear the grating roar
> Of pebbles which the waves draw back, and fling,
> At their return, up the high strand,
> Begin, and cease, and then again begin,
> With tremulous cadence slow, and bring
> The eternal note of sadness in.

But suddenly the sea becomes metaphor:

> The Sea of Faith
> Was once, too, at the full, and round earth's shore
> Lay like the folds of a bright girdle furled.
> But now I only hear
> Its melancholy, long, withdrawing roar,
> Retreating, to the breath
> Of the night-wind, down the vast edges drear
> And naked shingles of the world.

Bardsey could now be envisioned as a stubborn survival of venerable and transcendental values in the receding sea of religious enchantment.

At this moment, when so many values of the urban, industrial enlightenment were being challenged, Enlli also became the 'Island of the Dead' and the 'Wild Island'. The great theological debates of the mid-nineteenth century had questioned whether traditional visions of the afterlife – particularly eternal punishment – were compatible with modern standards of ethics and rationality. With its soil devoted to the keeping of 20,000 saints, Enlli became the ideal place to interrogate the posthumous persistence of the soul.

At the same moment another trend contributed to Enlli's celebrity: in the late nineteenth century English culture took a renewed turn towards the rural, seeking places the metropolis had failed to reach. Thomas Hardy's 'Wessex' was one product of this. So was the trend for literary figures to seek a simple life on Peak District or Cotswold smallholdings or in writing huts on the edge of their allotment. New societies, from the National Trust to the National Footpath Preservation Society, institutionalised the invention of the country-side. The garden-cities movement aimed 'to wipe the ugly nineteenth century from the slate' and show that the Industrial Revolution had been a bad dream from which Britain could awake. In their rejection of the present, journalists flooded west to find a past and future utopia.

Many of the flurry of Enlli articles that resulted are glorious in their vividness but horrifying in a chatty condescension that was amplified with every mile urbane travellers moved from the leather armchairs of their wood-panelled clubs. Charles Edwardes began 'The Island in the Currents' by describing the long Llyn Peninsula that stretches towards the holy isle.[3] He described hedged fields of fat little horses and black cows, punctuated by tiny grey villages and garish hills of purple and gold. He described kitchens filled with willow-pattern crockery 'of the kind extravagant and careless Saxons smashed out of use half a century or more ago'. As he travelled, he met 'broad-beamed housewives' and 'milkmaids, as sturdy as ever about the ankles' who 'continue to sing to their kine in Welsh'. The *Daily Mail*, he said, 'has bored into their midst and the scholars of the land exult in [this region]'. But he believed this to have had no impact on a people he saw as entirely provincial.

Edwardes stayed at the Ship Inn, the most modern of Aberdaron's three 'stumpy small' hostelries and 'the high-water mark of [the town's] splendour'. He found it full of the usual 'china monstrosities which to the Welsh of the Lleyn are objects of art', with chairs so spartan and outdated they 'would please the Spanish Inquisition'. He presents his room less as inn than ship, describing carpetless floors in bedrooms 'like the cabins of a schooner for size' where it is unnecessary to blow out the candle because the sea wind will soon extinguish it for you. In an augur of what is to come on the island itself, he describes how the churchyard (before the building of the long wall that now guards it from the sea) was subject to spring tides that 'pare away the graves . . . and play with skulls as it plays with pebbles on the beach'.

The eventual journey to the 'impossible island' takes place on a calm day yet is terrifyingly rough, through tides, waves and swell whose 'Atlantic quality' Edwardes makes great play of. He had no love for the sea: tides and waves are not points of interest but trials to overcome. The coastline is also an ordeal: a stinking strip of rotting weed and carcasses that is the last line of defence encircling a utopia. Within lie perfumed meadows filled with farms whose livestock, oats and potatoes make them 'as self-supporting as any in the Midlands'. The houses, Edwardes marvels, would be villa residences in London, worth anything from £30 to £40 a year in rent, yet here the rent is a mere trifle despite unique advantages including swathes of fine free pasture and unrestricted fishing in waters burgeoning with whiting, lobster and crab. No islet, he says, 'could make a more prosperous show to the casual stranger'; a crofter from the Hebrides set ashore on Enlli 'would think he was in a colony of lairds'. Soon Edwardes meets the islanders. They are capable women and strong-shouldered men, some 'with genuine Celtic red to their locks'.

A slow old man, with much grizzled hair to his head and his chin, and the signs of recent breakfast about his mouth, came towards us with a scythe. 'I am the king,' he said quietly . . . and then with a deferential little bow, he went his way to cut grass.

On the island Edwardes stayed in a room of 'the usual marine atmosphere', complete with stuffed puffin and colossal brass-bound Bible.

Only after pages of agricultural and economic exposition does Edwardes begin to recount his interviews with islanders concerning the theme which was quickly becoming the island's distinguishing feature for the outside world. The 'modern sceptic', he begins, surely doubts the story that 20,000 saints are buried here, but the experience of those who work 'this preposterous charnel pit' belies incredulous assumptions:

> It is all bones underneath, nothing but bones. I have seen them myself, indeed. There were womans with hair eighteen inches long, and childs, and mans, in such heaps as you could not believe, – no, indeed, except you saw them yourself. And their teeth, – oh, indeed, I never did see such full mouths of them.

These crops of teeth are mentioned in every article on the island; indeed Edwardes is rare in neglecting to draw specific lessons from them. Another, from the same year, takes their gleaming condition to show that these people 'lived more after the order of nature than the moderns' and pairs this idea with the observation that the islanders are free from 'those sanitary and other officials who vex plain functionaries on the mainland'. In all such articles, the inland city is associated with restriction, artifice and secularism, the tide-bound island with freedom and transcendence.

These general articles ran alongside 1890s texts that made the bones of saints their central purpose. In 'My Visit to the Island of the Dead', M. Dinorben Griffith presents 'a tiny Arcadic kingdom' in which the finest crops are fertilised by the multitudes of saintly dead.[4] 'Bardsey's present is peculiar, and, in some particulars unique,' insists Griffith, 'but the little island's past is so weird, not to say sensational, that the present is tame by comparison.' Whenever a new home is built, 'barrows full of bones' are dug up from trenches where the dead were laid head to toe in long lines. While birds 'sing of the ecstasy of life', 'the moaning of the waves' provides a requiem for the disinterred multitude. This journalist's vision of Enlli, defined by resistance to modernity and the embedding of the present in a long, uncanny past, persisted through the early twentieth century.

Only in the 1950s did new developments occur. The ultra-modern phenomenon of global warfare in the 1940s was a direct influence on the birth of island literature in general, and immediately preceded the creation of a literary Enlli built on total rejection of the city and as antidote to technocratic catastrophe. These writers didn't so much invent new meanings for the island, but transmutated 1890s themes from journalistic fragments into fully developed literary and artistic tropes. In the process, however, the dead that rise from the Enlli earth took a turn for the macabre.

The first major Enlli work of literature makes the island a particularly haunted, tortured place. Brenda Chamberlain lived here from 1947 and published *Tide Races* (1962) in the last year of her time there. Part memoir, part place-writing and part unfocused cry of anguish, this long prose poem is one of the most troubled and troubling books in the history of island literature. Gradually, the bodies rising from Enlli ground become its key motif: 'they get into our skins, the islanders, after they have shed theirs'. Enlli itself is a vast human body, its rocks sporting veins thick as human arteries:

The stone would seem to be composed of petrified tissues, skin, muscle, delicate bones. We ran our fingers over the filigree patterns. Falling to our knees we touched the remains of our ancestors. Or

302

their sculptured memorials in stone; their ivory-bright, bird-bone perfection, the metamorphosed flesh.

To touch this rock is not to feel the reassurance of a warm companion; it is chilling contact with a vast cadaver. Death is always present. When she visits the ruined abbey Chamberlain imagines not the peaceful worship that others might, but 'a plain-chant from perished throats', from dead who 'felt too keenly the cold austerity of outer regions of space'.

At around the same time, the poet R. S. Thomas was priest at the Aberdaron church that faces the island, having returned to the Welsh-speaking regions he called the 'repository of the condescension / of time'. When he crossed the water he found an island only a little less forbidding than Chamberlain's. A man in lifelong tension with the modern world, which he called 'the machine', Thomas traversed 'the gallery of the frightened faces of the long-drowned' to an island he presented as other-worldly and beyond the reach of time itself. This was before the 1970s resurgence of Welsh culture and language. Thomas prematurely mourned the passing of both and fixed on Bardsey, home to a single family rather than the ninety-strong community of its heyday, as a distillation of what had died:

> Here there is no sleep
> for the dead, they are resurrected
> to mourn. Everywhere is the sad
> chorus of an old people, waking to weep.[5]

The spectre of his absentee God haunted his experience of this sacred space:

> And God said, I will build a church here
> And cause this people to worship me,

And afflict them with poverty and sickness
In return for centuries of hard work
And patience.[6]

The slim volume on the island that Thomas penned with Peter Hope Jones expresses this loss in the bluntest terms: 'Bardsey's cultural image is a disappointment.' This is partly because of the broken thread of its population and partly because of the arrival in the eighteenth century of the Christian denomination Thomas found least congenial. 'The dank shroud of Calvinism', Jones and Thomas write, 'enveloped the island'. In the late-Victorian chapel, Thomas nonetheless found fellow feeling with saints who, a thousand years before, had also arrived too late to witness their God in action. The island's dark, inexplicable deity was already gone: modernity, and the triumph of 'the machine' over meaning, had for Thomas begun in the second century AD.

It is easy enough to see why this island might inspire works that are so wounded. Day after day I sat on the storm-ravaged coastline, watching the tides turn sea to scar tissue (figure 11.2). I rented a small stone cottage for my week ashore, without electricity or running water. Sitting through squalls, listening in darkness to walls of rain assault the thin windows, I found Edwardes' description of island 'villas' impossible to comprehend. I watched the island's one fisherman, Ernest, bounce in his small boat through a sea that was all aggression and wondered how he'd survived a lifetime in these elements. I found corpses of seabirds and watched the number of chicks that trailed behind a shellduck decrease gradually from five to zero. It was obvious how the island's effect on those traumatised by modern life might not be wholly uplifting. This identity, defined by death and dominated by the distant past, is too great a weight for any real place to bear.

It has been left to current islanders to reclaim the island from its past. Enlli remains a site of pilgrimage and its identity as the island of the dead, visited by hundreds of the would-be-holy, is unlikely ever to be shaken. Yet in different ways each member of the island's small community is invested in showing it to be a generative place, overflowing not with death but life, and productive in ways beyond accommodating visitors' needs. Not decay and winter storms but springtime and harvest define its new image. Bardsey, as much as Quandale or Ness, is thus a key site at which the difficult histories of coastlines are transmuted into visions of a productive and distinctive island future.

The island community is like the cast of a play. Each family has a clearly differentiated role, a niche in the island economy and a purpose in association with one of Enlli's aspects. The young Porter family runs the farm and leads the recreational use of the sea, with kayaks, and even the sky, with their hang-glider.[7] In photography and painting the family's two university-age children represent the island to the outside world, and with their ecological and naturalist expertise form a complement to the young volunteer ornithologists who stream through the observatory every spring and autumn. The observatory itself is a father-and-son affair, founded in the 1970s in recognition of the island's situation as a station on migration routes. Further along the dirt track, past the small chapel where a visiting preacher is often in residence, the warden and her partner oversee the island's organisation. As well as maintaining the smooth running of the cottages they organise the island's social life, corralling visitors into sing-songs and organising talks at the observatory.

This accounts for all the island families but one: the most long-standing island residents and the core figures in visions of island pasts and futures. This is Christine Evans, her husband (the fisherman and farmer Ernest), and their son, Colin. Each in their different ways lives a life in productive tension with the island's received meanings. Ernest is the strongest link back to the island's past, more comfortable in Welsh than English. Colin is the boatman who brings visitors to and from the island, carrying thousands of people and crates of food each summer. But he is the one figure most determined to give the island a sustainable life beyond its tourist income. In 2017 he bought the grounds and buildings that surround the lighthouse, planning a distillery to guarantee the island productive status for the future.

It is Christine, however, who is the presiding spirit of the once and future Bardsey. Small, sharp and energetic, she writes poetry that is, like the prose of Tim Robinson, both clear-eyed and full of wonder. Her verse and prose describes her relationship with the island, from her arrival in 1964 as a young tutor to island children, through her marriage to Ernest, into her present role as a kind of island laureate.[8] She strikes an acrobatic balance, taking seriously the island past but subtly revising the meanings it once had. Open and optimistic, she never dictates a specific idea of what the island is, instead exploring its diverse possibilities, finding her inspiration not when the island matches its archetype but when it generates surprise.

Yet there are certain threads that hold her work together. The most dramatic is yet another island embodiment. Her island and its seas are not cadaver, grave or charnel pit but, as Pippa Marland pointed out in the most perceptive study ever to feature Evans' work, its womb and crib:[9]

> Under the swollen moon
> Its body throbs warm between
> Water's shining wings.

The first poem of Evans' collection *Island of Dark Horses* (1995) begins in her warm, walled garden on this rock 'in an amniotic sea'. In the next, the ocean 'is a vast breathing cradle'. The third recounts returning to the island from the mainland, stepping ashore with 'birth-wet, wind-red faces' to a world in which 'the breeze / lets smells of growing things settle and grow warm'. The feet she depicts moving through small fields make 'new paths' rather than, like Thomas, retracing saintly footprints. Waves 'flower' around the feet, they are 'a trance of blossoming without rot, curdling without sour'. The shoreline is

frontier country: to walk here is to feel perception quicken and the intellect for once sits humbly with its dials gone dark; to seem immense as cumulus, infinitesimal as sandgrains in the suck of each wave's gathering.

Evans takes a small child to the shore, searching for cowrie shells, and sees in her 'fragments of the chrysalis I was' as well as imagining 'the wide-gazed woman this little girl will turn into'. This is an extraordinary poetry of growth, experience, memory and intimacy in which intensely personal scales of time and space are entangled with the evocation of thirteen centuries.

One of the richest poems in this collection recounts a thick summer mist that envelops the island, leaving islanders swaddled and enclosed as if in soft white bandages. The foghorn blares its colossal warning for five whole days and nights.

Timing it makes the moment
Momentous, makes me recall
Contractions. And, after each
A glistening bag of quiet sounds
Opens, drying off around us.

In the gaps between the bellows, some hear the mist as silence, and sense it as stillness, but stasis is illusory: 'all the smells of growing unfold moistly around us'. Evans refers to a sudden gap in the cloud as a tunnel like a bone-lined hole dug in Enlli earth, but then corrects herself, restoring the language of growth and nurture: 'not a tunnel:

a cocoon'. This poem's purpose speaks directly to those immensities each little place is wrapped in:

> What brinks, what late-summer vistas
> We are all ripening towards
> As we wait to see, wait
> For the sun to burn a way through.

In Evans' metaphors the many faces of life – marine and pastoral, domestic and wild – constantly intrude on one another, much as in *The Birlinn of Clanranald*. The island's tendency to erupt in unforeseen fecundity means nothing holds to its allotted place, culture and nature becoming indistinguishable. Lamplight after supper is soft as the sheen of buttercups, while the shadows it creates 'blossom' on the walls. The sea is a great heart pumping tides: in Evans' language all is pulse, seepage and flux. This is not a world without threats or death but one in which the greater, unquenchable force is always life. The wildflowers growing through the empty ribs of saints don't receive their meanings from the bones, instead the bones have their meaning remade by growth.

Evans' poetry echoes Enlli's images as seen in the 1890s and the aftermath of war, but all such meanings are refracted through her greater intimacy with the place. Her work gives the impression of respect for Enlli's past interlocutors, but bewilderment at their failure to see beyond their favoured tropes and actually observe the island in all its diversity. With delicious irony, she echoes the words R. S. Thomas wrote when he visited Bardsey looking for God. He was a mainlander travelling to Enlli, where she is an adoptive islander travelling ashore and seeking Ernest:

> Now, we trail alone round Marks and Spencer
> grow irritable searching;

all too often, pass on the wrong side
– I was there, but later –

The outdated crockery mocked by Victorian journalists has its parallel in her loving use of chipped ceramics and old plastic plates that 'make all our island meals an expedition'. Discarded by a mainland visitor a generation ago they are stained with blackberries picked by friends once stranded here by storms, the memories in these stains are island histories, records of many rhythms of sea-girt life. But these poems are also rich with histories. In 'Island Children' she recalls every Bardsey child since 1929, noting their characteristics. There is Billy Mark, boat-born, 'fearless crawler between cart-horse hooves', and four children at the Cristin farm 'cheekbones folded high like wings'. And she laments the lack of foresight that led to the island's emptying:

> 1870: fourteen families
> told to choose: stone harbour
> or chapel, larger, more devoutly distant
> from the main track of farmwork, gossip,
> washing spread across the gorse to dry.
> Trained to acceptance on this island
> that keeps its back against the dawn
> how could they see into a time
> it was no longer God's will to have grown
> two strong men from every house, to launch
> and row the sort of boat they knew
> or picture their trim patchwork fields
> rough as prairies, homes inaccessible
> as America? Asked, in turn, to speak
> how could they suddenly develop
> defiant strategies of choice?

In the long title sequence of the collection, all this overspilling of meaning, all these themes that clash without contradiction, are given explicit form. In italicised stanzas, monks observe the passing stations, but between their prayers the same sites are 'Rachel's kitchen' or the spot where Arthur picks caterpillars off cabbages. Evans notes the many ways in which past and present converge: 'a thousand years of monks' Latin . . . re-echoes' in the litany of species names for those caterpillars and the seabirds perching on the gable.

This is Evans' skill: to infiltrate venerable histories with the intimate present in ways that strip the past of pomposity while never diminishing its dignity. There is no other poet of islandness who captures so consummately the range of a place's implications, the transience and diversity of its meanings, or the foolishness of ever expecting a coastline to conform to the expectations we place on it. Her coast is a place to be experienced, not conceptualised, comprehensible only by beginning from intimate encounter. She is as sceptical as Norman MacCaig of 'big words', as critical as Tim Robinson of the mystical, but more open-minded than either to the ways in which every engagement with the shoreline is an enrichment of its accumulated meanings.

Leaving Bardsey was a wrench, particularly because it marked a dramatic change of tone. The greatest difference between kayaking the winter Atlantic and the summer Irish Sea was that what was coast became seaside. I had passed beautiful sandy expanses, from Sandwood Bay near Cape Wrath to Ballynahich Strand in Donegal, but these had not been places of ice cream and chips or even of buckets and spades. A mere two miles from the tides of Bardsey the gentle beach of Aberdaron was already filling with people as I launched at 8 a.m. A young brother and sister perched on the edge of a sit-on-top kayak, peering into water still enough for them to see the seabed; the beautiful sea gardens nearby, filled with shoals of fish and huge gangling spider crabs, are a snorkeller's paradise. A greyhound pelted down the strand, quickly a quarter-mile from a worried-looking owner: perhaps the only real spurt of energy this sleepy spot would see all day.

Small bays between rocky cliffscapes soon gave way to wide estuaries and long flat sands backed by mountains. Between the peaks and the sea lies land that has been more fundamentally reformed by humans than exposed Atlantic coastlines. Swathes of land that were once sea are now farmland. This gives a strange effect to much of the shoreline: low coasts as flat as meadows are contrasted by sea cliffs a mile or more inland. Tremadog crags are a climbers' paradise with a busy road not a writhing sea at their base. Cliff-edge Harlech Castle, at which Owain Glyndwr, Welsh history's greatest freedom fighter, held out against the English Crown, stands not above protective water but over caravan parks and a coastal railway that constantly restocks each beach with visitors.

Cardigan Bay was quick and easy work, showing few elements of Atlanticness: no islands, little raggedness and a sea that easily forgets whatever winds have passed across it. The turn westwards to the islets of Pembrokeshire was like a transition back towards the rougher seascapes I'd moved along for months. But these islands are without communities, their images dominated by the wildlife and heritage interests who own them and by the memory of their famous occupants in the golden era of archipelagic literature.

I'd intended to absorb myself in the world of 'the islandman' Ronald Lockley, spending a night on the islands, Skokholm and Skomer, that had been his world. But I was instantly met on Skomer by a representative of the Wildlife Trust of South and West Wales who wouldn't allow me ashore even for a few minutes without paying a £15 landing fee and, even when I had, made certain I was quickly on my way. Only in a place where familiarity with sea travel and respect for the water has been entirely lost could such attitudes prevail. Pushing quickly past Lockley country I was soon among headlands whose names sound more Cornish than Welsh and whose heavy tourist presence prefigured my final month of travel.

CORNWALL
(July)

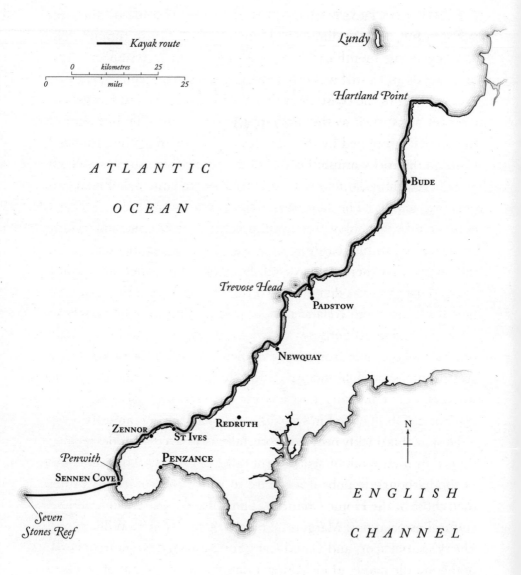

Kayak route

0   kilometres   25
0   miles   25

Lundy

Hartland Point

•Bude

ATLANTIC

OCEAN

Trevose Head

Padstow

Newquay

Zennor   Redruth
St Ives

Penwith   Penzance

Sennen Cove

Seven
Stones Reef

N

ENGLISH

CHANNEL

THE CONTRAST BETWEEN the truly wild life of Shetland in July and the diminished fauna of the Cornish summer felt, as I began this month's journey, nothing less than dystopian. With grey clouds and a stiff westerly whitening seas to a peculiarly deathly green, I held to the coastline rather than braving the last bastions of Cornish auks, such as the offshore isle of Lundy. The brutality of the sea felt simplified by this absence of visible life. When I moved through the rocky amphitheatres of the north Cornish coast, which elsewhere would clamour with auk, fulmar and gannet, I often heard only sea sounds. The fate of the chough illustrates this damaged ecosystem. A symbol of Cornwall, it is the bird still most intimately associated with this coastline yet it was driven to extinction here in the 1940s (that there's always still hope on interconnected Atlantic coasts, however, was demonstrated by the arrival of a tiny group of breeding pairs from Ireland in 2001). My feelings might have been different had conditions permitted me to land at the many stony beaches, inaccessible from the land, beneath the cliffs. Only able to cruise onto sand I found myself woken in the mornings by dog walkers, and launching and landing through lines of surfers and swimmers who bobbed in bays like emaciated neoprene seals.

Despite this thickly peopled shore, talk was scarce: attitudes seemed shaped by metropolitan detachment rather than coastal openness and in the harbourside pubs it was rare to hear Cornish accents rather than those of the Home Counties. The villages I passed all had their cultural associations: Margaret Thatcher's golf course, William and Harry's social spots and David Cameron's favourite restaurants (one of the middle names of his younger daughter is even that of a village on this coast). England's only Atlantic enclave felt more English than Atlantic, the coastline seeming as distant from a world of fishing communities as it was possible to be. Indeed, I met two speakers of the Cornish language during my journey. One was in Edinburgh, the other in Aberystwyth. Each felt they belonged to Cornwall but neither chose to live there. Today's leading singer in the Cornish tradition, Gwenno Saunders, lives in Cardiff and works for Radio Cymru.

At night, I was reminded of Westray as I listened to cattle bellow, but felt metropolitan when every landing led me to cafés and bars. I'd return to my belongings to find them moved and even, on Perranporth beach, a sandy red-faced child weeing on and in the kayak. Yet between the surf spots the water itself is not the leisure-sea of Wales; its Atlantic nature is a boisterous deterrent to flat-sea effluvia like jet skis and dinghies. The sea felt abstracted from the land: there was little criss-crossing of human and animal life to soften the shoreline. Indeed, this ocean abstraction would haunt me all month.

For the first two days, despite a harsh twinge in my left shoulder, I simply paddled single-mindedly onwards, wrestling the pallid swell with everything I could muster and asking myself, repeatedly, what on earth I thought I was doing thrashing across this strange coastline. Because I stayed some distance out, avoiding the confusion of swell reflected off granite, the little cliffs of Trevose Head rarely rose above the waves and I hardly saw the chunks of sea rock, stripped of the birds that some are named after, which litter the inshore waters. In this goal-oriented mode I took barely a photograph.

Only on the third day did the wind drop. And suddenly, reaching a different stretch of shoreline, my mood changed with my surroundings. I clicked into sympathy with the Cornish coast: the very abstraction of this green sea became the source of its resonance. I found myself in regions associated not with Margaret Thatcher but with some of the great Atlanticists: writers and artists from Virginia Woolf to Peter Lanyon whose engagement with this coastline has defined the modern meanings of the ocean. I became particularly absorbed by the post-war painters, poets and thinkers who – from Cornish homes in harbours or on clifftops – used the sea to reimagine the relationships of humans to their environment. Such reimagining had become an urgent cause when mid-century revelations of the extreme violence of which everyone is capable transformed perceptions of human nature forever. Through the work of thinkers who collected in 1950s Cornwall the Atlantic infiltrated new visions of what it meant to be human. New theories of movement, space and time used sea travel to explore the nature of existence in an irrevocably damaged world.

One of Cornwall's cliff-dwellers in the wake of war was perhaps the most thoroughgoing Atlanticist of all British poets: W. S. Graham. Where other coastal poets have been tied to specific locales – Vagaland to Shetland, MacCaig to Assynt, Evans to Bardsey – Graham's life lapped the length of the western seaboard. Born in the Clydeside town of Greenock, he spent much of his misdirected youth exploring the Highlands and Islands, developing the taste for rugged coasts that became the aesthetic and philosophical core of his art. He was a passionate but foolish lover of the outdoors whose jaunts to wild places sometimes flirted with catastrophe:

> I walked through Glencoe thinking I would get a lift not knowing the pass blocked at the other end and went unconscious on Rannoch Moor – was found in early morning by a shepherd who took me back to his croft and gave me brandy. Snow everywhere – had no coat or food – had been in open air for three days sleeping out with 8d in pocket. Before I became unconscious sitting huddled in a

sheepfold I imagined I heard girls laughing, that foxes came right up to me in the dark, that avalanches were starting. I wasn't frightened at all but enjoyed a lovely drugged kind of richness sitting in the numbing darkness – bloody pitch-black.[1]

Displaying lifelong distaste for convention and compulsion, Graham evaded conscription by seeing out the war on the west coast of Ireland (his mother was from Galway); he thus arrived in 1940s Cornwall on an Atlantic trajectory not dissimilar to mine. He is the kind of poet whose prickly gregariousness, hands-on practicality, and fondness for strong drink might lead him to be called 'earthy' were it not a different element that soaked him through. Reading Graham's essays, letters and poetry reveals a waterlogged imagination: stripping him of his watery metaphors would leave him almost mute. He wrote two famous long poems about the sea: *The White Threshold* (1949) and *The Nightfishing* (1955). But those are really poems about human nature: to Graham the salient characteristic of the human self is that it is uncontainable, inconstant and unknowable as ocean. *The White Threshold*, he wrote, is 'the most sea poem for a while I've seen. And more "human" than anything I've ever made.'

In contrast, many of his poems may seem to be about human life, but are formed from abstraction and fluidity inspired by sea and rivers. Obsessed with Joyce he took the notion of 'stream of consciousness' more literally than others did; fascinated by Heraclitus, the idea of existence as constant flow pervades his work. Heraclitus' river that can never be stepped in twice became Graham's metaphor for life. And death, at the end of the river, is ocean:

> When we are with the night-sea round our feet
> And feel the music of green-weeded shores,
> And even when the salt death-breeze of night
> Is on our faces with impatient breath,

Then we will turn and look along the way
That we have come, along the shining stream,
Right to its misty birth.

It is often impossible to tell which – sea or self – is Graham's subject and which metaphor. Water and identity disperse each into the other, the poet and the person lost and reborn in the deep.

In these ways, the sea was not just a subject for Graham's poetry but its medium. Aiming to 'unsettle' the language of the sea, he used the sea to unsettle language. He praised friends, poets and artists in terms of their tempestuous potential to unsettle him. His letters are full of passages that express this unsettling: 'shall I compare thee to a day of storms. Batter my Gurnard's Head. The wind roars and the waves rumble in . . .' There is no bracken in these letters except 'drenched bracken', there are no hands in his poems but 'saltcut hands'. His metaphors for creativity saw poems 'swimming up' from the deeps. He used the amphetamine Benzedrine (alongside copious liquor) to charge his pen, calling it 'my spare engine in this bull-shouldering sea' of life. And others took on his metaphors when talking about Graham's work: the poet and critic George Barker compared Graham's output to the sturdy creations of Greenock shipwrights that could proudly sail the world.

When he arrived in Cornwall Graham moved into a caravan, perched on sea cliffs, which he conceptualised more as a ship than a home. He named it 'the wheelhouse', after the pilot's station on a boat, and 'the arkvan' because seas seemed to flood up towards it. And he liked to assure his correspondents of its place at the locus of elements:

The gales from the sea rock me about and my poor arkvan, and flood comes in through the wrenched seams. I'll have to be fixing scuppers round the walls to drain away the waters. Three days now of a constant gale coming up over the shore and lifting sand and salt. The rooks are blown about and the fulmars sail well inland. I've a new highnecked navyblack jersey so the cold can't get me. Only there's wind and wet, winter and angry weather. The air has a smell of iodine and drowned fields. Water water wish you well.

Deep in poverty (Graham was one of the few poets of his era who refused to make ends meet by earning money by non-poetic means), he grew potatoes, snared rabbits and scoured the cliffs and hedgerows for sustenance. He was described by visitors as living on nettle soup, wearing trousers 'worn thin as pyjamas', and begging for boots to keep his 'poor iambic feet off the flint of Cornwall'.[2] Later, he moved to a cottage on the cliffs at Zennor, where he would do occasional shifts as a voluntary coastguard, spending long days in the clifftop lookout hut writing poems (there was even one occasion when he was alert enough to save a life). Like so many Atlanticists he preferred to write in darkness – 'awake and alone in the middle of the night in a little igloo of lamplight' – and in uneasy silence, where 'I could not entirely forget the weather'.

Graham's move to Cornwall expressed his rejection of Scottish efforts to create a national poetic style and he always viewed linguistic revivals as little more than conscriptive artifice. On a much reduced scale, Cornwall faced the same conundrum as Scotland in this era, seemingly forced to choose between championing the literature and traditions of a threatened ancient language or promoting a rich dialect linked to an industrial working class. In both Greenock and Cornwall, Graham refused to make such choices, preferring to listen to language as it was used rather than express preferences on how it ought to be. But despite this rejection of the Scottish 'scene' he pined for Scotland's coastline, seeming always on the verge of a move to the Western Isles. The home of his imagination was a collage of islands and long shores. His imagined sands were imprinted with the 'arrow steps' of gulls, and filled with sight and sounds of coastal waders: 'The curlew's cry travelling / still kills me fairly'.

There is something about the long poems of the mid-twentieth century, steeped in history, nature, myth and war that, despite their torque towards dusty abstraction, is uniquely moving and profound in the quest for personal reconciliation with deranged times: Eliot's *Four Quartets* (1943), Lynette Roberts's *Gods with Stainless Ears* (1951), David Jones's *The Anathemata* (1952), Iain Crichton Smith's *Deer on the High Hills* (1962), Basil Bunting's *Briggflatts* (1966) and H. D.'s *Winter Love* (1972) all achieve this damaged grandeur.[3] Most are tied to specific places, and rivers and sea run through the work of these middle modernists as through the writing of few other generations. Seas are symbols of destruction, threat and loss, but also of revitalisation, tranquillity and desire. Rather than its all-too-easy associations of permanence the sea meant flux: both devastating frenzy and glittering hope for a generation who saw all that was solid melt into water.

These images could operate on the largest scale: world wars were storms that left numb, becalmed, seascapes in their wake. But personal emotion was also an ocean: when Bunting evoked his passion for a distant lover, storm clouds, roaring seas and unquellable tidal energies became the turbulence and vigour of physical love. The hostility of the sea which 'perfects our loneliness' becomes his image for the tumult of passion parted from its object. The result was the weird intense ode 'I am agog for foam' (1926). All these poets (and others, such as Pound and Auden) exploited the symbolic power of the sea, but Graham was unusual in making it the central image of his major works and a consistent influence on his method: *The White Threshold* and *The Nightfishing* are the only long poems of the post-war decade to set their search for human meaning entirely at sea.

For Graham and his contemporaries in Cornwall, the sense of loss that resulted from war was compounded by an awareness of the passing of coastal lifeways. Graham drank – prodigiously – in old Cornish pubs, as he had in Greenock. Breaking into song at the drop of a hat he made Cornish friends (and enemies) quickly, and was soon invited out to sea in search of herring. Not yet knowing his Greenock heritage, the fishermen he worked with were surprised at this poet's accomplished sea legs. He insisted that the long nights in pubs on the harbour were, in fact, his work. On such occasions, 'the muse herself is drunk' and the 'shapes of language . . . spill round our ears'. He committed the cadences 'of drinking and affection' to notebooks 'so that later I might explore the mechanics of their memo-rableness and vitality'. The poet must, Graham insisted, build an organic rhetoric to charge up the formal elements of poetry until its energy is amassed and released like a breaking wave.

Obsessed with 'disturbances to language', Graham lived among people whose whole lexicon was sinking to oblivion. His response was expressed through mixed metaphors of animal and water: 'the language is a changing creature continually being killed off, added to and changed like a river over its changing speakers'. The Cornish artist Andrew Lanyon made a series of scrapbooks – evocations of Cornish life in image and prose – several of which are now held in the Cornish Studies Centre at Redruth.[4] These are the most revealing and poetic descriptions imaginable of the world into which Graham entered. Lanyon's father, the artist Peter Lanyon, was a friend of Graham born in the same year, and Andrew characterised their world in terms of a shift in maritime life: they awoke 'to a world in love with coal and observed the dying flame of sail'. The wars had spurred massive development of seafaring technologies and one unforeseen effect of this was a dramatic linguistic watershed. The words of Cornish boatbuilding, pilchard fishing and sea craft spilled round Graham's ears but are alien to ours.

Lanyon argued that the written collections of folklorists who descended on Cornwall en masse in the twentieth century are a limited form of salvage. He took the example of 'Horensheboree': the syllables fishermen sounded as they dragged a boat across the shore and the noise used to co-ordinate a tug-of-war (it is, the dialect collector might say, a kind of Cornish 'ready steady go'). From the folklorists' collections the syllabic tangle remains mysterious:

> we can only assume its rhythm and guess where the essential pause might occur. In fact the 'ee' followed the rest after a second or two delay, before the next heave . . . The power of the native version lies partly in where the sounds are made – deep in the throat, near to the ground. 'Ready steady go' is suddenly too sporty and dislocated from reality, elevated into the stiff upper lip of competitiveness rather than the chanting of togetherness.

For all its artifice and refraction of reality, poetry like Graham's could convey rhythms and cadence of seafaring that prose could not.

The communal traditions such language was tied to occurred on exceptional scales. When pilchards arrived in summer, thousands of men and women wielded three-ton nets that each stretched a quarter of a mile round the shoals. Almost a billion of these fish were transported annually from the four main Cornish ports during the eighteenth-century heyday of the phenomenon. I was surprised to learn that – after being piled into five-feet-high salted walls that wound like glinting rivers through every Cornish harbour – most were shipped to Italy. In the same era, tin and copper mines were extended deep under the Atlantic. Miners worked the cavernous galleries to the sound of thunderous rumbling as ocean undertows dragged huge boulders along the seabed. In the face of storms, Graham wrote, 'the stones roll out to shelter in the sea'. Many alluvial tin mines operated only in the winter because they required fast-flowing streams to operate, so miners were freed to either fish or smuggle through the summer. Smuggling and wrecking became perennial, if subsequently exaggerated, associations of the coast: the sea could seem to wash things clean of ownership even while import duties were ever more widely policed. When good fortune aligned for all these industries, the wealth that poured in could be extravagant. But none were stable nor consistent and a single decade of dearth could have an impact on Cornish culture as great as a century of plenty.

The decades around 1900 saw many forms of dearth. The last 'native speaker' of the Cornish language died in 1891. The pilchard fishing reached the end of its long decline, and the last 'Huers' posted on hills to watch for shoals were retired in 1922. Harbours suffered neglect: St Agnes crumbled into the sea in 1916 and was never restored. The period has even been labelled 'the Great Paralysis' by one leading scholar of modern Cornwall.

At the same time, however, cultural self-consciousness was growing. Attempts to revive Cornish began around 1900. Gorsedh Kernow, a tradition of bardic festivals mirroring the Welsh Gorsedd, was founded in 1928 and was running strong during Graham's Cornwall years, even if still a Welsh-driven statement of pan-Celtic identity rather than something independently Cornish. After the end of the language as a widespread vernacular, linguistic culture remained not just distinctive but enormously varied. According to Andrew Lanyon, a dialect line splits even the small Penwith Peninsula as if a 'different people lived either side of the spine'. Vowels, like the weathered rock, he says, were eroded more on the seaward side perhaps by the influence of Breton and Irish. The language was thus full of 'Nuggets of Norse and alluvial Erse' as 'the great rolling gait of the seafaring mother tongue swallows small fry without a hiccup'. This was the fragmentation and flux in which Graham began to interrogate the crises of self and society that the wake of two wars entailed.

Every word of the Cornish masterpiece that is *The Nightfishing* is metaphor. *The Nightfishing* lays not just a single journey on the line, but a whole life, an entire civilisation, and the condition, universal to humans, of being constrained by the very language that permits expression. The central question the poem poses concerns the elusive nature of the self – the detachment of every person from their inner being. But the questions it posed for me as I travelled were why the Atlantic was so powerful an expression of this theme and why this use of the ocean should issue from Cornwall. Though interwar British modernism was run through with coastal influences, these were largely the stuff of sheltered North Sea or Irish Sea resorts (Margate, Aldeburgh, Swanage or Morecambe); they were rarely the less hospitable Atlantic.

On the evening of the day I entered Graham country I reached St Ives. The pretty streets were heaving with people and the narrow passageways of the old Sloop Inn almost impassable. I settled in a corner where I could watch the evening pass. Families sat playing board games amid arrays of bottles and glasses that made each table like a chess set. This town was where Graham came to meet people who wished to be 'isolated together', and it was in the work of Peter Lanyon, one of his companions in mutual isolation, that I felt the answer to my questions could be found.

The principle of being 'isolated together' is telling of St Ives' history. The only real harbour in north-west Cornwall, the town has long been integrated with the other outposts of the Atlantic edge. This sheltered inlet linked Wales, Brittany, Iberia, the Channel Islands and Ireland (so that it is said to have been the first place in Britain to receive Guinness). Latin, French, English and Cornish were spoken by medieval Ivesians, while Welsh, Breton and Portuguese have also been staples of its streets. The sea saints came to this region too, scattering chapels across the peninsula and linking the literate cultures of Cornwall, Wales, Ireland and the Hebrides.

But this was a place transformed in 1870 by the arrival of the

Great Western Railway. Suddenly, a town of working fishermen and women became a shard of metropolis lodged in the coastline – a champagne cocktail spilled in the Atlantic. Perspective shifted from vistas on a vast Atlantic world, to inward-looking Englishness. The railway's ethos fostered an essentially imperial relationship between centre and periphery, encouraging an image of Cornwall as different, primitive and exotic. The great photographers of the age – names like Preston, Frith, Valentine and Moody – visited and sold their wares in the metropolis. St Ives had two cinemas in an era when film going was a distinctively urban pursuit (no town in the world of St Ives' size could say the same). The young Virginia Woolf and others of her set arrived for holidays. Woolf watched dolphins pass from the harbour and the image of 'a fin passing far out' became a strange cryptic motif that runs through her writing. Her memory of St Ives inflected her characters' experience of London, the sound of tube trains passing becoming a regular rumbling of ocean waves. For Thomas Hardy, who arrived in Cornwall the very year the railway opened, this coastline stood for aspects of English identity that smug metropolitans dangerously chose to repress. The whole structure of his Cornish stories, *A Pair of Blue Eyes* (1873) and 'A Mere Interlude' (1885), is underlain by weird new temporalities. Everything is ancient: vegetables and flowers from the Scilly Isles are unloaded at Cornish ports just 'as in the time of the Phoenicians'. But his plots would be impossible without the rapid rail travel between London and Penzance that compressed into a few hours what had till 1870 been a week's journey. Tourism is the only industry he seems to notice.

But it was painters who were soon the most famous St Ives phenomenon. They recreated picturesque and rustic harbour sights and rushed them for sale to the city. Scenes of traditional life were gathering new meanings amid twin perceptions that the quality of urban life was in decline and that urban life in general was physically and morally corrupting of its subjects. Artists brought visitors in their wake: the seascape painter Borlase Smart, for instance, designed posters to advertise the Great Western Railway to escapist London crowds.

In this era, when artists still painted with nineteenth-century assumptions concerning what reality was and how it should be repre-sented, the relationships between artists and fishing community were friendly and mutually supportive. There were only occasional clashes, such as the time an artist was thrown in the harbour for painting on a Sunday. But Peter Lanyon was a misfit in this scene. Born in St Ives, he might seem a natural bridge between the two populations: an insider to the fishing community where other artists were outside observers. But embracing elements of abstraction – 'blobs and smears' as some locals saw it – he instead contributed to the end of the uneasy alliance. The fisher community's respect for artists had been built on the idea that they were skilled craftspeople representing the town, realistically, to the outside world; they couldn't see themselves, nor any technical craft, reflected back from modernist abstractions.

Lanyon grew up in the wealthier part of St Ives (upalong), on the hill above the harbour (downalong). Atlantic rollers that hit the coast were dispersed, by the time they reached upalong, into pointillist mist that coated windowpanes and abstracted the seascape below. In the nursery, his sister would begin to tell stories that Peter would subvert, filling them 'with storm clouds'. 'He was', she said, 'always one to side with the wind.' People and landscapes were similarly subverted in his paintings, everything generalised from an initial specific vision. A place or person was painted over and over till any specificity was just suggestion, and human bodies could be headlands or headlands bodies. Black lines weave round his paintings. They are 'beachcombed', in his words, from the black nets that lined the St Ives harbour. The key to understanding his artistic project is that lines like these do not concern space so much as time: they are temporal elements to guide the eye round the canvas, analogous to the eye following a wave as it rises and unfurls. Smooth curves lead vision smoothly and swiftly, but tangles and complexities are obstructions like reefs in water.

Lanyon was formed by war more directly than Graham. It was his immersion in combat, and in the Libyan desert where he fought, that turned him from a figurative artist into something new. He insisted that revelations of wartime technological genocide posed a challenge that demanded from the artist perspectives other than the urban; he began to identify the new toughness war and its aftermath instilled in him with the hard granite of his home. Where other modernists made technology their subject, Lanyon pursued similarly radical objectives through materials such as sea cliffs that were interpreted by others as timeless and permanent. Post-war people were strangers to the familiar, he said, and if art focused only on superficial elements of radical change it would neglect the human transformation underlying them. Who now could truly believe, he asked, that cities or technologies were solutions? Writing to a German critic he insisted that 'it is in the bare places like West Cornwall . . . that many artists will find an answer for their times'.[5]

Not just the bare rock, but the physical and spiritual challenge of the indifferent ocean drove his quest for post-war meaning. It was when watching waves break outside Tripoli that Lanyon had become obsessed with the question of how to represent movement on a static sheet of paper. The art historian Rosalind Krauss has referred to the sea as a special kind of medium for modernism because of 'its opening onto a visual plenitude that is somehow heightened and pure, both a limitless expanse and a sameness, flattening into nothing'.[6] In similar vein, Lanyon wrote that 'the effort to understand and to live with and to adjust to vastness calls out an equal depth in our own psyche creating anxiety enough to trigger off a rescue operation'.

This sense of the viewer's disorienting immersion in earth and water was crucial to Lanyon's art, suggested perhaps by St Ives artists' outside view of the worlds they painted and his own zealously felt sense of Cornish insider status. A sufferer of vertigo, he would run to a clifftop and aim to recreate in paint the experience of the shocking edge rather than to represent the view below. Like a cubist he painted in ways that evaded the singular viewpoint, aiming to see things as if from many angles. All is deconstructed. There is, it is often said, only air and never sky in his paintings, and only rock, never cliffs. Yet somehow – to me at least – the formulations usually used for Lanyon's elements don't fit his seas. They are always ocean, never water, in part because it is their massed movements that matter to him: there is spray, foam, wave and, as in the painting *Headland* (1948), the shimmer of light on oceanic expanse. Given how much is written about Lanyon it's strange how little is written about his sea. His modes of transport on land and in the air – from motor-biking to gliding – have been subject to intense scrutiny, yet boats are rarely mentioned among his methods, despite his frequent seafaring and the number of his famous paintings, such as *Godrevy Lighthouse* (1949) or *Silent Coast* (1957), that specify boat-bound perspectives or watery movement. It is perhaps precisely because the sea is unique among his elements that Lanyon scholarship has so stubbornly refused to mention it.

The results of Lanyon's method were described by many viewers as abstract but by Lanyon as landscape. Abstraction, he insisted was merely a tool for rendering land and sea as experiences not as objects. Some critics sided with the fishermen in dismissing Lanyon's work as meaningless. But John Berger, the great scourge of abstract art, was convinced of the constructive and embodied nature of Lanyon's project:

> he searches for something which includes a sailor's knowledge of the coastline, a poacher's knowledge of the cover, a miner's knowledge

of the seams, a surveyor's knowledge of the contours, a native's knowledge of the local ghosts, a painter's knowledge of the light.

This life in a no man's land between figurative and abstract art might perhaps explain why his seas are different: water is, in its real state, already more abstract, less concrete and more transitory, than land. Perhaps this is why modernists from Woolf to Graham to Maxwell Davies were so eager to embrace it. Lanyon's abstraction is no more than a wave that flows across reality. As the undertow draws back, familiar elements of boat, cliff, seabird or human appear, though still refracted by the veil of sea that thinly covers them.

By 1950, when Graham and Lanyon would drink together in St Ives, they were each as obsessed as the other by the question of boundaries between self and world. Lanyon's painting interrogates edges and his prose is littered with phrases that muddy distinctions between landscape elements and human bodies: 'I would not be surprised if all my painting now will be done on an edge – where the land meets the sea where flesh touches at the lips'; 'I like to paint places where solids and fluids come together, such as the meeting of sea and cliff, of wind and rock, of human body and water.' In most artists it would be safe to assume that quotes like these referred to multiple paintings in different genres, but for Lanyon each quote could describe a single brushstroke. His nudes, in his words, evoke 'the sensation of oneself against impossible weather . . . it is information about certain sensuous qualities that I'm after . . . transformed to an understanding of landscape'. These nudes are as likely to make visual reference to Cornish headlands as to particular parts of a human body, just as there are suggestions of human form in his landscapes: an echo of the body is 'something a viewer subliminally responds to and seems to take part in'. Indeed, even Lanyon's demonstrative painting practice muddied these human/landscape divides, swift sweeps of a heavily loaded brush across canvas were fitting metaphors for the fluid forces of nature meeting cliffs. The point was that the human body and its environment are bound into a single structure. The body defines its own boundaries through movement in the landscape and every perception of the external world is linked to a bodily reaction which generates new perceptions: 'if bodily space and external space form a practical system . . . it is clearly in action that the spatiality of our body is brought into being'. In these ways the painting, operating in all four dimensions, is as much an act of self-definition against the immensities of ocean as is *The Nightfishing*.

Rosalind Krauss speaks of the sea as a medium of modernism 'because of its perfect isolation, its detachment from the social'. But this is not Lanyon's Atlantic, nor Graham's, since the point for both of them was to humanise seas: to see them as layered with social events and activity. They presented landscape visions that were no longer stripped of centuries of occupants but that achieved their meanings through use. As he moved along the coastline, painting his responses to what he saw and felt, Lanyon considered lives lost in sea disasters and mining catastrophes (such as the occasion in 1919 when the 'man engine' in the Levant mine collapsed, killing thirty-one):

At present I seem to be on a pilgrimage from inside the ground, as if I were the only one saved from the Levant disaster as if I moved, an unlucky mourner, along the gale-ridden coast to St Just . . . Therefore due to my own failings my work contains the whole constructive process which I illustrate as follows: the miner extracts inside the earth; his trolleyings in the galleries, a shuttling within the earth and his laborious incisions are eventually brought to grass. Here the change continues by controlled processes in the furnaces and eventually the product has no resemblance to the rock ore. That is the mechanics of it.

As all these evocations might suggest, Lanyon was a fierce Cornish nationalist. His name at the Gorsedd of Bards was 'the Rider of the Winds' in reference not just to his haunting of the clifftops but his occasional method of viewing the coast from a glider. He insisted that his work was characteristic of the Cornish people – insular but seafaring, provincial but transnational, practical but dreamy. Indeed, there is actually as much of romantic Cornwall in Lanyon's art as in any episode of *Poldark*; the techniques and media might be diametrically opposed, yet the aesthetics are remarkably consonant. Lanyon's deep commitment to his own small Atlantic corner was perhaps the biggest difference between his and Graham's attitudes to the sea. All else seemed to run like parallel lines through otherwise messy waters of the post-war world. Confrontation with the Cornish coastline was, as Lanyon insisted in a 1964 lecture (just weeks before his death in a gliding accident), the existential challenge in which the new human nature, demanding revision in response to the horrors of the Holocaust, could be constructed:

> I believe that landscape, the outside world of things and events larger than ourselves, is the proper place to find our deepest meanings . . . landscape painting is not a provincial activity as it is thought to be by many in the US but a true ambition like the mountaineer who cannot see a mountain without wishing to climb it or a glider pilot who cannot see the clouds without feeling the lift inside them. These things take us into the places where our trial is with forces greater than ourselves, where skill and training and courage combine to make us transcend our ordinary lives.[7]

I stayed two nights among the winding stone passages and galleries of St Ives, travelling inland twice to read in the Cornish Studies Centre. The winds rose, fell and rose again as I eyed up my moment to embark on the final jaunt of my journey. Eventually, I launched in bright silver light into a fresh morning breeze. Wind and waves soon subsided enough for me to wander through rock archways and into caves, weaving between reefs and enjoying the cold of the glistening sea foam. Finally, a porpoise broke the surface and, once the tide dropped low, I found myself in technicolour sea gardens of plumose anemones, jewel anemones and hydroids. Grappling with Lanyon and Graham seemed to have opened my eyes to this coastline in ways I hadn't thought possible in those grim first days at sea. By the early afternoon I'd reached Sennen Cove – the last long strand before Land's End – and I had a choice to make.

I weighed up three options. The riskiest and most demanding would be to aim for the Scilly Isles thirty miles offshore. The changing weather made this option hard to sanction. The simplest and safest alternative would be to hug the coast south-east towards the Lizard and Penzance. But it was a third option that intrigued me most and for which conditions seemed strangely suited: I chose to end my journey in the mythical land of Lyonesse. This would allow my final voyage to echo the circumstances of its conception in the Summer Isles, with an escape to abandoned islets in the middle of the sea.

Lyonesse is the core of mythic Cornwall: the land of Tristan and Isolde and a staple throughout the Celtic imaginary of Europe. It was an idea beloved of Victorian sages such as Tennyson and Swinburne and of later Cornish revivalists such as James Dryden Hosken whose poems were collected as *The Shores of Lyonesse* (1923). The archaeologist O. G. S. Crawford conveyed the essence of this Cornish Atlantis:

ONCE upon a time (so tradition says) a region of extreme fertility lay between the Scilly Islands and Cornwall. This land was called Lyonesse; and where now roll the waters of the Atlantic there once stood prosperous towns and no less than a hundred and forty churches. The rocks called the Seven Stones, seven miles west of Land's End, are said to mark the site of a large city. This country was overwhelmed by the sea, and the sole survivor, one Trevilian, escaped destruction only by mounting a swift horse and fleeing to the mainland.[8]

This fabrication is not altogether without basis. In the eras from which this region's many Neolithic remains date, the Scilly Isles (or the single Scilly Isle as it may still have been even in Roman times) was a far larger land mass than today, plausibly a seat of significant power. There are remains nearby of submerged forests from which beechnuts can still sometimes be found, and the lore of fishermen speaks of large chunks of masonry emerging in nets cast near Seven Stones. This dramatic reef is violently abrased by sea and above water for only a few hours at low tide but was, in an age of lower waters, once a larger structure. This is thus a region whose ferocious elements have long been filtered through myth:

> The shout and vision of the sea gods grey,
> Stampeding by the lone Scillonian isles.

The Lyonesse myth held huge attraction for Cornish writers because its tales of Tristan and King Mark, Isabelle, Iseult, Melodias ap Ffelig and King Hoel Mawr depict a unified Celtic culture, before the advent of the nation state, when Cornwall was in constant inter-action with Brittany, Wales and Ireland. By the 1930s Cornish Revival culture had gone wild for Lyonesse and Arthur, and the Seven Stones had become a symbol of past freedom.

My good fortune today was that a spell of still air, before the rising breezes, coincided with tides that reached low ebb shortly after midnight. As I set out, conditions felt like a compendium of much I'd seen so far, from my initial journey to Cleirich onwards. But the glinting silver light on the slightest swell recalled most closely my Skellig voyage. Then I'd been on my way to rocks successfully inhabited for 600 years before abandonment, riddled with built structures. Now I was embarking for a place whose only inhabitants since prehistory have been the ghosts of 200 shipwrecks. In the lighthouse era Seven Stones had been judged too dangerous to ignore but too challenging a site for even the Stevensons to attempt to construct a light. The unusual solution was a huge floating lighthouse moored off the rocks in 1841. Manned by a crew of five, this light-ship must have been the most terrifying place to live. It was frequently ripped from its moorings and found drifting into skerries, or limping to the Scilly Isle port of New Grimsby. Indeed, the early history of this strange vessel is filled with drama and disaster: drownings, collisions, wartime bombings and even the dramatic explosion of a meteor above the deck. After multiple enforced replacements, the ship is now a vast red monstrosity – the highest contrast with this vivid blue green sea – visible for miles around.

The reef, when I arrived at dusk, was chaos: a two-mile-long eruption in the calm. I'd naively assumed the water would break on one side of these rocks, providing the hope of a sheltered landing. But the reef was less like a wall in front of waves and more like a knife blade thrust into ocean, violence on its every edge. Landing was impossible. I was forced, for the first time in my life, to overnight on the water. The hours of darkness were brief, but the focus required to stay safe – listening for every danger – made them seem to last forever. I found myself in cold sweats as I watched the many lights by which I could roughly assess my position. I thought constantly of the galloping Trevelyan as I forced my tired arms to keep me moving. And my relief was intense when lighthouse beams began to dim against the earliest hints of morning. First light, when it finally arrived, was gorgeous beyond compare, the transition from cold blue to rich gold intensified by a dark weather front approaching from the south. As the sun emerged, warm winds stirred. I downed the pint of milk in my dayhatch before pressing weary limbs into the last fifteen-mile slog of the journey. As on the day I'd reached Cleirich, winds rose as I neared my goal, and arriving at the cliffs before the weather truly changed was a respite I could not have done without. This time, however, I was returning to land, perhaps never again to occupy the water as I had since that Cleirich nightfall.

The journey to Seven Stones had been among the only abortive efforts of my coastal year: one of the few times I'd met with water whose movement proved an uncrossable barrier. Yet there was nothing at all remarkable in the conditions. My strange night meandering through swell in darkness drove home the extraordinary good fortune I'd met with. Later, at home, I explored Met Office visualisations of a year that had seemed to exist under a meteorological charm. These were freak seasons when autumn gales were somehow delayed till January, when rainfall as well as wind speeds had been a tiny fraction of their usual levels. Autumn had been like summer and winter like spring. What I'd gleaned from my journey along these unusually placid Atlantic coasts was just a taste of the wildness and a hint of the power of the most colossal force against which this little island group rubs up. I'd witnessed communities facing the fair-weather ocean and heard their tales of weather-bound hardship, but gained less sense than I'd expected of what it meant to live with the sea's rage day after day. All this had left me, however, ready to begin the task of considering how the nature of Britain is changed by the view from the sea.

THE VIEW FROM THE SEA

I T WAS A whole year after reaching the Seven Stones that I sat down to write this epilogue. I'd barely been in a kayak, or even seen the sea, since. Instead, I'd sat in the fourth floor of a redbrick building in the heart of England, teaching, writing and administrating. I felt less embedded than ever before in Peter Lanyon's larger powers that define our movement through the world or Tim Robinson's 'immensities in which each little place is wrapped'. Even the screech of peregrines that nest in the clock tower by my office did little to conjure connection to the Atlantic elements I'd left behind.

What I missed most was immersion in constant movement: the world view from low in the wave. I missed the sense of being part of a vast, coherent dynamism. Indoors I was sometimes unsettled (a condition I could only refer to as 'the bends', since it was caused by coming up from the sea) and sometimes resorted to a sleeping bag in the garden among the foxes and green woodpeckers. Never before had I so welcomed rain: a good cold soaking was the best medicine of all.

As well as feeling the contrast between two sensory worlds, I found myself more attentive than before to certain characteristics of the urban society I'd returned to. The managerial ethos of large modern institutions, even universities, is one in which standardisation is assumed to be a universal good and any independence or eccentricity must be fought for. I prickled anew at every attempt to 'standardise procedures' and felt, intensely, the contrast between my urban life and the power of locality I'd witnessed in communities I'd travelled through. There, other ways of life had survived only because of awkwardly independent souls from Sorley MacLean to Annie MacSween.

So I decided in the autumn to return to Ireland. I threw myself back into the writhing life round the Blasket Islands and meandered along the ragged bays and headlands of the Dingle Peninsula. I sat once again, a mile offshore from stupendous cliffs, in vast swell and spray through which dolphins arced and gannets dived. On the first morning, I overturned my boat to submerge myself in sea and watch a pod of dolphins pass; I spent the afternoon with shivering body but ignited, elated mind. I paddled on, musing on Robinson's Aran dolphins: so perfectly commensurate with the wave that they prompt the viewer to ask what it would mean for humans to fit so neatly into the world. Returning here had been a strategy to prompt ideas about the difference that seeing Britain and Ireland from their watery edges makes.

While travelling, I'd seen repeatedly that traditional narratives of British and Irish history cannot possibly account for the composition of these coastlines. Those interpretations aren't just landlocked but largely London-locked, imposing shapes on the past that explain how the metropolis became the place it is. If my sole goal was to understand the Atlantic half of Britain, I thought, I might as well throw out all the histories of Britain that line my shelves. In them, periods of urban, inland ascendancy are read, implicitly, as progressive, and other geographic perspectives have little place. One result is that the whole long process of historical change is seen in terms that imply not just the clearing of watery impediments to modern life, but the human conquest over nature, to be a universal good.

In these ways, our historical narratives are not just incomplete stories about the British and Irish islands; they are ethically unsuited to today's world. The eradication from our historical vocabularies of large-scale labels such as Renaissance, Enlightenment and Modernity, as well as agricultural and industrial revolutions, can allow us to begin to reframe our histories in less urban and goal-oriented ways. In their place can be built narratives that show the frictions: the geographical unevenness, the pain and the visceral resistance to two centuries of rapid change that were once envisaged as heroic progress but are recognised as more problematic with every year that passes. Where speakers of Scottish Gaelic, Irish or Welsh were once told to abandon their language with phrases like 'we have to move on, we have to be modern', the kind of modernity those instigations envisaged is now itself seen less as economic necessity than ecological catastrophe.[1] Recovering the voices of the small-scale fisherman, the coastal crofter and the leaders of threatened communities is among the best methods we have of building histories that aren't triumphalist propaganda for the events that created the problems of our present.

Not just narrative labels but geographic formations merit rereading. I'd had no conception before beginning this journey that the Hanoverian suppression of the Highlands, the great famine, the Education Acts and the actions of Shell off Erris in the 2010s were all part of the same process in which diverse geographic concerns have been disciplined out of national life, with the wealth of the urban, inland centre subsuming and consuming not just all resources but all attention.

We live with the consequences of the eras that pushed integration and improvement as an ideal. There is still widespread confusion concerning what Britain is or means. Politicians, journalists and even historians write as though Britain is a naturalised entity that has existed for many centuries. Yet books with 'Britain' in the title still often feature only England or even London. Even movements such as 'four-nations history' usually remedy this by integrating Edinburgh, Cardiff and Belfast into their interpretations. The nation becomes a network of capitals, defined more than ever by its cities.

There is no doubt that the ideal of an integrated urban Britain has been significant at certain historic moments, even when its cost has been high. When the process of building Britain as an imagined unity was at its height, the nation was led into the First World War by a Welshman whose first language wasn't English, while the 'national church' (as the Anglican communion was still labelled) was led by a Scottish archbishop of Canterbury. Victory in that conflict is often presented as the moment of the great success of the idea of Britishness. Yet that was only half the story: this was also when the Victorian dream of Britain died. Irish resistance to integrative forces, and refusal to be coerced into a 'national' ideal of Britishness, is a reminder that there has never been anything inevitable about either the idea of Britain or the ideal of integration.[2]

Despite the comforting myths of an 'island nation', protected by seas that lap gently at its borders, wherever land meets ocean perspectives surge in that erode the expected boundaries of Britishness. There are few better experiences to prompt this disorientation than *really looking* at maps of the archipelago. To look closely is to recognise the vast, un-cited swathes that are nonetheless filled with communities, the long ribbons of coast whose first language isn't English, the islands whose inhabitants feel their difference and practical independence keenly, and the multitudinous points of connection to African, Iberian, Scandinavian and transatlantic ports. It is to see proportions shift: the busy fractals of the west slow the eye (like Lanyon's net-inspired tangles) in contrast to the short easy sweeps of south and east.

Those fractals are parts of the oceanic geographies that Barry Cunliffe recovered from scholarly obscurity in *Facing the Ocean: The Atlantic and its Peoples, 8000 BC to AD 1500*. Yet that important text, and the debate in its aftermath, had one unhelpful element. The Atlantic worlds Cunliffe identified were not substantially undermined in the fifteenth century, as the chronological bounds of his study suggested, but only in the nineteenth.[3] In the larger scheme of things, it has been just the blink of an eye since the Atlantic edge was these islands' centre. And the geographies that marginalised the coastlines and moved life inland have themselves been exceeded by the decentring influence of new technologies. As the reintegration of island crafts and coastal industries continues, thanks to e-commerce and communication, there need no longer be centre and margins, nor severe geographic inequalities, in quite the way there was before.

But still more dissolves at the edges than traditional historical narratives and accepted geographies: the very subject matter of history is challenged too. The ocean, as an engine of climate and a vast ecosystem interlocked with air and land, is at last being taken seriously in our societies, thanks to TV presenters more than politicians; yet the question of what that means for our histories is only beginning to be asked. The book series Oceanic Histories, begun at the University of Cambridge in 2018, promises to be one new vehicle for the necessary analysis of world oceans, while the publication of Helen Rozwadowski's *Vast Expanses: A History of the Oceans* (2018) at around the same time makes this an exciting moment for oceanic histories.[4] Philosophers increasingly demand a downgrading of the traditional, Kantian and Cartesian, status of human beings, arguing that the human mind is not an ultimate arbiter of meaning but that 'things' – including oceans, whales or algae – have meaning independent of what humans think of them; yet mainstream histories have found few ways to give non-human animals or things more than supporting roles in their analyses.[5] Artists, such as the freediver Janeanne Gilchrist, create exhibits to challenge the idea of human control and self-possession. In Gilchrist's photographs, plastic bags are suspended in shifting ocean and thus appear animate: though the intended purpose of these objects, she insists, lasts only for moments, they take on lives of their own once discarded and live long afterlives outside of their relation with humans. With a few notable exceptions, however, historians still write as though humans controlled the things they construct. Histories thus have blind spots that mirror the gaps left by urban outlooks on the present: they, too often, assume historical processes end at the boundaries of human society.[6]

To begin to observe the past from the water is, of course, a tiny step in the grand scheme of rethinking histories to speak to a world in which crises and concerns can't always be addressed within familiar frameworks. But coastal pasts allow us to recognise the intertwining of human and ocean worlds as well as the many trajectories that have led our communities to this fractured present.[7] They are thus good places to start.

In making such a start, the histories of these islands' edges emerge not only as the catastrophic events, such as famine and clearance, which are most commonly commemorated in coastal museums. Those events were brief, though dramatic and significant, disruptions to rich and continuous histories packed with vigour and success. This is where the literature that has played so large a part in this book comes in. The reams of historic poetry, song and fiction that have poured from the edges, and the languages in which they've been spoken, sung and written reveal the textures of those living cultures and run the full emotional range from joy to farce to tragedy. The poetry of Rob Donn, Alasdair mac Mhaighstir Alasdair, or Màiri Mhòr nan Òran is not a memorial to decline but an inheritance for an era in which it has begun to resonate more widely. From the 1750s to the 1970s, Atlantic regions were painted as belonging to a bygone world; as the bishop of Ramsbury in Wiltshire put it in 1976, the west coast of Scotland is an 'area of decline, an area where everyone's thoughts are of past glory and present doom. There is said to be a Highland problem, and no one yet has found the solution.' This 'Highland problem' had been an invention of the Scottish Enlightenment used to muster enthusiasm for the spread of 'improved' farming. The grim irony is that, when invented, it had been wrong; but the new standardised agriculture instigated the famine and depopulation which amplified a thousandfold the perceived need for cities to 'save' the coastal 'victims' of history.

It seems rarely to have occurred to inland-dwellers before the late twentieth century that the problem might lie not in the islands and among the mountains, but in the Lowland cities whose policies excluded them from new geographies of mass communication and consumption, though some lovers of the islands, such as Alastair Dunnett and Seumas Adam, pressed the case. Improvement and centralisation compromised flourishing Atlantic networks without providing infrastructure through which Highlands and Islands might take part in the new commercial society on a level footing.

Only in the last few decades (particularly with the creation of the University of the Highlands and Islands) have these perspectives been voiced in scholarship. Jim Hunter's most profound work, *The Other Side of Sorrow* (1995), remains the finest evocation of Gaelic literary sensibilities and ecological consciousness through the modernising era; David Lloyd's *Irish Times* (2008) is the best manual for how to read the trajectories of Atlantic histories anew; and scholars from Meg Bateman on Skye to Tim Robinson in Connemara mediate coastal literature to large new publics in many languages.

Today, all the languages of the edges are growing, though slowly. For the first time in a century populations are increasing too. There is a new fascination with the local, and a new awareness of the threats that tourism can pose to fragile communities and ecosystems. The 'sea ethic' that Rachel Carson called for in *Under the Sea Wind* (1941), founded on a fish-eye perspective on the ecosystem, is finally a public concern that can nudge society's gaze towards the edges.[8] At the same time, the cultures of cities are recognised as drastically unsustainable. The result is that the past presentation of islands as 'victims', and cities as 'saviours', is finally beginning to be reversed. Forward-looking uses of the past in Ness, in contemporary crofting, and in fishers' knowledge projects might all be small initiatives, but they represent the basis for something big. Each of them answers Lloyd's call not to lay the dead to rest and consign them to a bygone world, but to reinvigorate past lives and outlooks as part of a conscientious route into the future. In this context, writing history doesn't entail the reconstruction by rational moderns of particular long-dead pasts, but rather the imaginative exploration of unrealised past potentials.[9]

And the coastlines that the dominant cultures of the nineteenth and twentieth centuries marginalised are full of such potentials. They are places where the centralised political economies never worked. In the face of attempts to universalise an ill-fitting urban culture, all Atlantic coastlines suffered and many societies died. But some survived, or even flourished through their suffering. These were places in which inspirational individuals proved capable of rousing communities in celebration of locality and littleness, against the homogeneity of nation, and in favour of the attentiveness to the immediate environmental realities that can universalise the experience of living. I finished this journey with a new resource – a new mental archive – of Atlantic thinkers whose ideas represent the wisdoms on the world I felt I'd begun to learn. These Atlanticists' heartfelt protests against the marginalisation of coastal cultures deserve far more anthologising, broadcasting and championing than they have ever, so far, had.

I wondered how much the journey had changed me otherwise and thought again of the ragged map of Britain whose every western indentation now conjured a story, an emotion or a physical sensation. I realised that immersion in these worlds had not, as I'd expected, cured me of my romanticism. I hadn't met, on the journey, many people who worked the sea whose attitude wasn't something like romanticism: awe before the ocean's power to give and take life, along with a fascination for the sea's wild and sensuous stories, seemed to be present on every front line of confrontation with the water. It isn't romanticism that needs to be cleared from perspectives on these places, but the assumption that these communities somehow belong to the past, not the future, and are merely hazy places to escape to.

The reason certain poets, musicians and artists came to matter to me as I travelled was the closeness of their observation, whether dealing with community, fauna, foliage or the more-than-rational interplay between human self and ocean. They came to represent the power of experience in forming the texture of ideas. I laughed out loud at certain lines of Christine Evans or W. S. Graham because I saw them echo things I knew to be real only because my odyssey had forced me to look closely at waves, plants and people. Far from instilling an alternative vision to the naive romanticism I set out with, the journey had shown me that a romanticism which delves into the natures of humans and their fellow species, finding wonder while rooted in the real, might not be so naive after all.

ACKNOWLEDGEMENTS

I T'S IMPOSSIBLE TO know how to approach the acknowledge-ments required in a project like this, since the whole thing would never have got going, and would have fallen apart many times thereafter, were it not for the kindness of multitudes of friends and strangers. My first thanks must go to the three people whose inter-ventions made the journey possible. Adam Nicolson saw my outdoors blog, invited me to the Shiant Isles, and suggested this project to his agent and editor: without him, the idea for the journey would never have come about. Georgina Capel, Adam's agent, then took me on board at Capel Associates and helped me conceptualise the book: her advice and support has been invaluable at every stage. Arabella Pike, editor at William Collins, showed extraordinary faith in my ability to undertake this journey and provided crucial, insightful guidance throughout, turning a wild idea into a completed project. I'd like to thank the people whose involvement allowed the book to take its final form: Christina Riley for designing the sea-fan section breaks, Martin Brown for the maps, Joe McLaren and Jo Walker for the cover design, and Donald S. Murray for allowing me to use his poem 'The Cragsman's Prayer'.

This book would also have been impossible without a supportive institution: few universities would have permitted me the time and freedom to write this text in the way the University of Birmingham did. Successive heads of research, of the history department, and of the school of history and cultures – Matthew Hilton, Corey Ross, Nick Crowson, Elaine Fulton and Sabine Lee – have my immense gratitude for the way they support staff through times that are, for all universities, extremely challenging. Several members of the University of Birmingham's Centre for Modern British Studies kindly read and commented on sections of the text: particular thanks to Chris Moores, Matt Houlbrook, Mo Moulton, Zoe Thomas and Jonathan Reinarz. Thanks too to wonderful Birmingham historians Simon Yarrow, Marga Small, Lucie Ryzova, Chris Callow and Beth Parkes. My students on modules such as Sites and Sources in Modern British Studies, History in Theory and Practice, and Reason and Romance also helped me formulate lots of the ideas contained in the book, even putting up with questions in history exams concerning salmon and ocean plastics. One further institution, the National University of Ireland, Galway, gave invaluable support to the project through a Hardiman Research Fellowship in the spring of 2018. Thanks especially to Nessa Cronin, Kevin Sutherland and Lillis Ó Laoire for making me feel so welcome in Galway and for introducing me to scholarship on the Irish Atlantic. And thanks, as always, to Peter Mandler: the best mentor a historian could ever have.

The other people who made the book possible in a practical sense are the hundreds of individuals who picked me up as a hitch-hiker, helped me carry kayaks or even took time out of their Valentine's Day evening to drag me out of a marsh I'd, extremely embarrassingly, got stuck in (thanks Mike and Anne Grabham, owners of Loch Seil Holiday Cottages). I wish I'd been better at recording the names of everyone who had helped in these ways. Thanks to the staff of the Dolphin Centre, Bromsgrove (now Bromsgrove Sport and Leisure) for training me up to be able to do any of the carrying and kayaking at all. Then there were the people who welcomed me for coffee, whisky or a bed for the night when I arrived at their homes, flats or even yachts, including the wonderful Porter family on Bardsey, the writer of exceptionally beautiful prose about Shetland, Sally Huband, the inspirational kayaker Nick Ray on Mull, the psychologist, historian and champion of Gaelic culture Finlay Macleod, the historian of west-coast culture, Jill de Fresne, Jim and Morag Hewitson on Papay, the poet and chronicler of Lewis and Shetland, Donald S. Murray, and the couple whose magical work has done more than anything else to bring these coastlines to literary life, Tim and Mairead Robinson. Others arranged days out for me; thanks to Eoin Warner for the seaweed foraging, and to the ecologist Paul Thompson and artist Neville Gabie for a trip to Eynhallow that was a particular highlight. Thanks to Peter Muir (Achiltibuie) and Jo Shaw (Shieldaig) for providing me with wonderful coastal places to spend weeks reading and planning the journey.

Several artists and poets gave me a great deal of help and inspiration, either in person or by correspondence. These included Moya Cannon, whose extraordinary generosity was invaluable during my Irish journey, Christine Evans, Peter Maxwell Davies (as well as his partner, Tim Morrison) and Errollyn Wallen. Several publishers, publications and broadcasters helped out by featuring the journey in one way or another. Thanks to BBC Breakfast TV, BBC Radio Orkney and Visit Orkney; thanks to Jordan Ogg, Sarah Laurenson and Mallachy Tallack at *The Island Review*, David Knowles and Sharon Blackie at *Earthlines*, Martin Bewick and Ella Johnston at Dunlin Press, and thanks to the history and literary journals *Past and Present* and *Green Letters*.

The help I received at archives, particularly on small islands, often went far beyond what would usually be expected of archivists or historians. Many of the people who run these archives are among the most inspirational I met: the staff of the Shetland Museum and Archive, Orkney Library and Archive, the Scottish Studies Centre, Annie MacSween in Ness, Mairi Ceit MacKinnon on Barra, as well as many others along the way such as Jane MacLachlan and Fiona Mackenzie.

There are several scholars blazing significant pathways in aspects of Atlantic history who were generous enough to let me read their work and without whom sections of this book could not have been written. Pippa Marland's work on island literature was truly inspirational and substantially informed the sections on Tim Robinson and Christine Evans. Glorious interviews conducted by the oral historian Erin Catriona Farley were similarly crucial. The exceptionally innovative Orkney archaeologists Dan Lee and Antonia Thomas also provided an array of ideas. Nessa Cronin's work was essential for the Irish sections of the text, and for informing my attitude to geography throughout. Others, including Philip Hoare, James Macdonald Lockhart, Anna NicGuaire, Isaac Land and John Gillis provided ideas, and responses to my ideas, that never failed to revitalise my enthusiasm.

There are a few friends and family with whom I've travelled in the past, or who it feels like I've been undertaking this journey with: Benjamin Thomas White, Sadiah Qureshi, David G. Cox, Diccon Cooper, Matthew Pritchard, Chris Hill, Valtteri Arstila, Anne, Gordon and Rosie Gange, Thelma and Valerie Wood, Mike, Gwyneth and Gruffudd Owen. But the biggest thanks of all goes to my partner, Llinos Elin Owen, with whom I took up kayaking (after a car accident in 2009 left her unable to walk in mountains any more); an inspirational kayaker, now training with the team GB Paralympic squad, Llinos pushed me to overcome fears of rough seas, and provided the perfect sounding board for every idea about kayaking or small-language communities that's in this book.

The author and publisher would like to acknowledge the following for permission to reproduce extracts: Birlinn Ltd for Norman MacCaig's poems 'Apparition', 'Boats', 'Non pareil', 'Patriot', 'Poems for Angus' and 'Wreck', poems copyright © the estate of Norman MacCaig, reproduced with permission of the Licensor through PLSclear; 'Northern Lights' from *The Collected Poems of George Mackay Brown* by George Mackay Brown, copyright © The Estate of George Mackay Brown 2005, reproduced by permission of John Murray Press, an imprint of Hodder and Stoughton Limited; 'North' from *North* by Seamus Heaney, courtesy Faber and Faber Ltd.

NOTES

INTRODUCTION

1 Keith Thomas, 'The Tools and the Job', *Times Literary Supplement* (1966).

SHETLAND

1 Rachel Carson, *The Edge of the Sea* (1955); the other two classic texts are *Under the Sea Wind* (1941) and *The Sea Around Us* (1951); in the light of current awakening to ocean crises, it seems likely that these texts will soon be recognised, alongside *Silent Spring* (1962), as central pillars of the most important ecological oeuvre of the twentieth century.

2 Bryan Nelson, *The Gannet* (2010).

3 George Morrison, *One Man's Lewis: A Lively View of a Lively Island* (1970).

4 Robin Robertson, 'The Law of the Island' (2010).

5 The book I spent the evening with was Alison Munro, *Small Boats of Shetland* (2012).

6 John Cumming, 'Namin a Boat' from *White Below: Poems and Stories from Shetland's Fishing Industry* (2010).

7 Cumming, *White Below*.

8 Like the gannet, Shetland seems to draw flights of intensified language from the scientists who study it, the leading geologist of Shetland, Derek Flinn, calling the islands 'erosional remnants standing proud of the North

Sea floor'; D. Flinn, 'The Coastline of Shetland' in R. Goodier (ed.), *Proceedings of the Nature Conservancy Council Symposium, Edinburgh* (1974).

9 Michael Grieve, 'Foreword' in L. Graham and B. Smith (eds), *MacDiarmid in Shetland* (1992).

10 Hugh MacDiarmid, 'On a Raised Beach' (1934).

11 Jen Hadfield, 'Butterwort' (2005).

12 Jen Hadfield, 'The Ambition' (2010).

13 Samuel Hibbert, *A Description of the Shetland Islands* (1822).

14 This is the poem that begins De Luca, Johnston, Sinclair & Wiseman (eds), *Havera: The Story of an Island* (2013) from which the place names, and stories of the Laurensons, below are also taken.

15 Goodlad adds a sad epilogue to her tale: 'So dey towt, an I believe dat, if dey hed lived anidder year in Havera, dey nivver widda left. Dey aa said dat becaas da motor boat cam eftir dat.

16 Gerald Bigelow, Michael Jones and Michael Retelle, 'The Little Ice Age, Blowing Sand and a Lost Township', *New Shetlander* (2007); George Low, *A Tour through Orkney and Schetland* (1774/1879). Thanks to Blair Bruce at Shetland Museum and Archive for directing me to these texts.

17 George Mackay Brown, *Northern Lights* (1999).

18 William Porteous, 'The Isles of Thule' (1894).

ORKNEY

1 Jim Hewitson, *Clinging to the Edge: Journals from an Orkney Island* (1996).

2 An excavation was conducted here in May/June 1990 by Chris Lowe. With very limited resources, Lowe used a tapestry technique to analyse representative fragments of the site. Almost everything he recorded will now have changed. His work was published as *St Boniface's Church, Orkney: Coastal Erosion and Archaeological Assessment* (1998).

3 W. P. L. Thomson, *The New History of Orkney* (2008).

4 John Cumming (ed.), *Working the Map: Islanders and a Changing Environment* (2015).

5 The most extensive analysis of the impact of maritime culture on gender relations has been conducted in relation to Shetland: Lynn Abrams, *Myth and Materiality in a Woman's World, 1800–2000* (2010).

6 The most rewarding of these recordings, from which many of the quotes in this chapter are drawn, were made by Kate Towsey and Helga Tulloch in the early 2000s. Extracts from their interviews were published in two wonderful collections, *Orkney and the Sea* (2004) and *Orkney and the Land* (2007).

7 Iain F. Anderson, *To Introduce the Orkneys and Shetlands* (1939).

8 Several recent studies analyse the interplay between different periods in the history of this region: Daniel H. J. Lee, 'Quandale: The Biography of a Landscape', masters thesis, University of the Highlands and Islands (2008); Daniel H. J. Lee, 'Northern Worldviews in Postmedieval Orkney: Toward a More Holistic Approach to Later Landscapes', *Historical Archaeology*, 49 (2015), 126–48; Matthew Butler, 'The Landscapes of Eynhallow', masters thesis, University of Bristol (2004); Antonia Thomas and James Moore, *Eynhallow, Orkney Survey 2007: Data Structure Report* (2008).

9 This term was coined by Tim Ingold in 'The Temporality of Landscape', *World Archaeology* (1993). It is now used widely by archaeologists and historians; see also Tim Ingold, *Being Alive: Essays on Movement, Knowledge and Description* (2011).

10 Tom Muir, 'The Creatures in the Mound' in *The Mermaid Bride* (1998).

11 Tim Robinson, *Connemara: The Last Pool of Darkness* (2006).

12 Justin Vickers, 'Peter Maxwell Davies's Variations on a Theme: A Catalogue of the "Sea" Works', *Notes* (2015); Nicholas Jones at Cardiff University is currently researching Max's uses of Orkney land and seascapes.

13 Finlay is surrounded by odd myths, including several about his Orkney years. Almost everyone who writes about Finlay and Rousay says that he was a shepherd there. As Alistair Peebles showed, this is simply not the case: when he came to Rousay to write he paid his way by taking part in

the enormous Orcadian road-building efforts of the 1950s. It is also often said that Finlay had been evacuated to Rousay during the war; this too is untrue.

14 Philip Hayward, 'Aquapelagos and Aquapelagic Assemblages', *Shima* (2012).

15 Brathwaite saw western philosophy as founded on a static conception of dialectics, reliant on fixity, assuredness and appropriation. He proposed in its place 'tidalectics': an oceanic alternative that favoured transitory fluidity. Gaston Bachelard had insisted that 'a being dedicated to water is a being in flux', and Brathwaite argued that the same is true of languages, societies and everything else; see Stefanie Hessler (ed.), *Tidalectics: Imagining an Oceanic Worldview through Art and Science* (2018) and Anna Reckin, 'Tidalectic Lectures: Kamau Brathwaite's Prose/Poetry as Sound-Space', *Anthurium* (2003).

THE WESTERN ISLES

1 Iain Crichton Smith, 'Between Sea and Moor', *Towards the Human* (1986).

2 Donald MacDonald, 'Lewis Shielings', *Review of Scottish Culture* (1994); see also a wonderful university architecture project by Catriona MacDonald, *Aig an Airigh* (2012).

3 Outer Hebrides crofting survey 1956–61.

4 See for instance Domhnall Uilleam Stiùbhart, 'Local traditions concerning the late medieval history and topography of Sgìre Nis' in Rachel Barrowman (ed.), *Dùn Èistean, Ness, Isle of Lewis: The Excavation of a Late Medieval Clan Stronghold* (2015).

5 Roger Hutchinson, *A Waxing Moon: The Modern Gaelic Revival* (2005).

6 The most substantial book ever published by the Ness Historical Society is a thoroughly researched and expansively illustrated account of Ness and the Great War: *At the Going Down of the Sun* (2016).

7 Hutchinson, *A Waxing Moon*.

8 Crichton Smith, 'Between Sea and Moor'.

9 Daniel Corkery, *Synge and Anglo-Irish Literature* (1931).

10 Hutchinson, *A Waxing Moon*.

11 William Chambers, 'The Gaelic Nuisance', *Chambers Journal* (1877).

12 Besides Hutchinson, Sharon Macdonald, *Reimagining Culture: Histories, Identities and the Gaelic Renaissance* (1997) also provides detailed coverage of these themes.

13 'Terrific Display', *Ness News* (1978).

14 'Revolution in a Gentle Way', *Community Care* (1981).

15 James Hunter, 'Who Says History is Bunk? The Past Inspires Ness to a Better Future', *Press and Journal* (1979); see also 'Revolution in a Gentle Way' as well as a host of other articles in the Ness archive.

16 Hutchinson, *A Waxing Moon*.

17 *Sinn Fhein a rinn e: Proiseact Muinntir nan Eilean, 1977–92*, Comunn Na Gaidhlig pamphlet, Lews Castle Archive, Stornoway.

18 This phrase, a rallying-cry of the 'history from below', pioneered by social historians in the 1960s, comes from the preface to E. P. Thompson's classic text, *The Making of the English Working Class* (1963). More recently it has been adapted to Highlands and Islands contexts, playing a pivotal role in Jim Hunter's (almost equally classic) *The Making of the Crofting Community* (1976) and cropping up repeatedly in archival documents relating to the Comunn movement.

19 Hunter, 'Who Says History is Bunk?'; Comunn Eachdraidh Nis was not alone for long: it led a wider island renaissance that germinated in the depths of a national depression. At the same time as the people of Ness developed their vision of a Gaelic historical revival, a few Skye idealists dreamed up a similar fusion of past and future. Slowly, a small Gaelic library, established in farm buildings at a coastal spot named Sabhal Mòr Ostaig, became something fiercely ambitious. The project was initially the eccentric-looking idea of a Gaelic-learning landowner, Iain Noble. In January 1975 Noble wondered whether it might be possible 'to start a course which would last either one or two years with some kind of award or diploma at the end. The courses would be in Gaelic, and would be intended either as an addition to or a substitute for a university degree. Students would be taught about the flourishing literature of the Gaels, local history

and environmental subjects of every kind including geology and biology. There might be scope for certain practical subjects as an optional extra, such as agriculture, trawling and other trades, and the theme could be to educate people for life in a rural community, rather than to train them for professions in the cities.' In this statement, the idea for the University of the Highlands and Islands was born.

20 Erskine Beveridge, *North Uist; its Archaeology and Topography* (1911); a wonderfully inventive instance of historical geography, partly fictionalised, tells the story of Beveridge's life on Vallay: Fraser MacDonald, 'The Ruins of Erskine Beveridge', *Transactions of the Institute of British Geographers* (2014).

21 See for instance Robert Macfarlane, 'A Counter-Desecration Phrasebook' in Di Robson and Gareth Evans (eds), *Towards Re-Enchantment: Place and Its Meanings* (2010); the phrase also now has official UNESCO usage.

22 For analysis of the 'Counter-Desecration Phrasebook' idea see Jos Smith, 'An Archipelagic Literature: Reframing the "New Nature Writing"', *Green Letters* (2013) and *The New Nature Writing: Rethinking the History of Place* (2015).

23 Kathleen Jamie, 'A Lone Enraptured Male', *London Review of Books* (2008).

24 The UK system of climbing grades consists of two parts. The first half, known as the 'adjectival grade', relates to the overall difficulty of the climb. The 'E' at the start of all the climbs listed here stands for 'Extremely Severe' and is the highest category of all; the numbers that follow (1–9) denote just how 'extremely severe' the overall climb is (lower categories, such as Hard Very Severe, aren't split in this way). The adjectival grade takes into account exposure (the fear factor) as well as the level of technique required and for that reason is highly subjective. The second half of the grade (the technical grade) relates only to the technical demands of the most difficult single move on a climb, denoting the level of skill required to achieve that manoeuvre (using a, b and c they run from 4a to 7b, although 7a and b are still quite rare). The Child of the Sea, at E5 6b is thus a truly terrifying prospect.

25 *Historia Norvegiae*, extract from A. O. Anderson, *Early Sources of Scottish History* (1922).

26 An early acknowledgement that this question should be debated is found in Murdo Macaulay, *Aspects of the Religious History of Lewis* (1984).

27 Dougie Maclean's 'Pabay Mor', with its description of a 'wild Atlantic son' embarking, via quiet Loch Ròg and windswept Gallen Head, on 'the shoulders of the ocean, on the bare back of the sea' is a fine example.

28 Donald Monro, *A Description of the Western Isles of Scotland called Hybrides* (1549).

29 B. E. Crawford (ed.), *The Papar in the North Atlantic* (2002).

30 This was the Papar Project, led by Barbara Crawford, which resulted from a conference held at St Andrews University in 2001.

SUTHERLAND AND ASSYNT

1 Frank Fraser Darling, *The New Naturalist Guide to the Highlands and Islands of Scotland* (1947).

2 Ian Grimble, *The World of Rob Donn* (1979).

3 Calum Maclean, *The Highlands* (1959).

4 Màiri Sìne Chaimbeul, 'The Sea as emotional landscape in Scottish Gaelic song', *Proceedings of the Harvard Celtic Colloquium* (2002).

5 Maclean, *Highlands*.

6 Fiona MacDonald, *Missions to the Gaels: Reformation and Counter Reformation in Ulster, Highlands and Islands, 1560–1760* (2006).

7 Grimble, *World of Rob Donn*.

8 Norman MacCaig, 'Apparition' (1988); see also Colin Nicholson, 'Such Clarity of Seeming: Norman MacCaig and his Poetry', *Studies in Scottish Literature* (1989).

9 Norman MacCaig, 'Backward Look' (1954).

10 Norman MacCaig, 'Patriot' (date not known).

11 Raymond Ross, 'Norman MacCaig: the History Man' in Joy Hendry and Raymond Ross (eds), *Norman MacCaig: Critical Essays* (1990).

12 Norman MacCaig, 'Old Maps and New' (1970).

13 Norman MacCaig, 'Wreck' (1950).

14 Norman MacCaig, 'Basking Shark' (1967).

15 Norman MacCaig, 'Neanderthal Man' (1980).

16 This is a sequence of twelve poems, published as 'Poems for Angus' in 1978; for a treatment of these poems as elegy, see Nathalie Ingrassia, 'Norman MacCaig and the Fascination of Existence', unpublished thesis, (2013).

A MOUNTAIN PASSAGE

1 G. M. Trevelyan, *Clio: A Muse, and Other Essays, Literary and Pedestrian* (1913).

2 R. W. Emerson, *English Traits* (1856); see also Robert Richardson, *Emerson: The Mind on Fire* (1995).

3 For recent treatment of Trevelyan and Youth Hostels, see Michael Cunningham, '"Pavements Grey of the Imprisoning City": The Articulation of a Pro-Rural and Anti-Urban Ideology in the Youth Hostels Association in the 1930s', *Literature and History* (2016). The rediscovery of Trevelyan runs in parallel to the echoes and returns identified by David Matless among geographers: 'Nature, the Modern and the Mystic: Tales from Early Twentieth Century Geography', *Transactions of the Institute of British Geographers* (1991).

4 A. R. B. Haldane, *The Drove Roads of Scotland* (1952).

5 John Dixon, *Gairloch in Wester Ross* (1886).

THE INNER SOUND AND SKYE

1 Graeme Warren, *Mesolithic Lives in Scotland* (2005).

2 See Gabriel Cooney, 'Seeing Land from the Sea', *World Archaeology* (2010); also Tim Phillips, 'Seascapes and Landscapes in Orkney and Northern Scotland' in the same edition of *World Archaeology*.

3 Hugh Carthy, *Burren Archaeology* (2011).

4 Norman MacCaig, 'Boats' (1953).

5 A wonderful essay by Meg Bateman, 'The Bard and the Birlinn' – in an excellent but sadly scarcely available collection, C. Dressler and D. U. Stiùbhart (eds), *Alexander MacDonald: Bard of the Gaelic Enlightenment* (2012) – explores the poem's metaphors in detail.

6 This phrase is from Alan Riach's literary rendering of Alasdair mac Mhaighstir Alasdair's *Birlinn Chlann Raghnaill: The Birlinn of Clanranald* (2015).

7 Kenneth Steven, 'The Small Giant' (2009).

8 Colin Simm, *Otters and Martens* (2004).

9 Richard Hugo, *The Right Madness on Skye* (1980).

10 W. H. Murray, *The Hebrides* (1966).

11 Malcolm Slesser, *The Island of Skye* (1981).

12 Sorley MacLean, *Traighean* ('Shores') (1943).

13 Emma Dymock, 'The Quest for Identity in Sorley MacLean's "An Cuilithionn": Journeying into Politics and Beyond', unpublished thesis (2008).

14 On the songs and poetry of this region, see Meg Bateman (ed.), *The Glendale Bards* (2014).

15 Sorley MacLean, *Ris a Bhruthaich: The Criticism and Prose Writings of Sorley Maclean* (1987).

ARGYLL AND ULSTER

1 Colin Nicholson, 'Poetry of Displacement: Sorley MacLean and his Writing', *Studies in Scottish Literature* (1987).

2 Joan Kelly-Gadol, 'Did Women have a Renaissance?' (1977).

3 Courtney Campbell, 'Four Fishermen, Orson Welles, and the Making of the Brazilian North East', *Past and Present* (2017).

4 Ninian Dunnett, 'Introduction' in Alastair Dunnett, *The Canoe Boys: The First Epic Scottish Sea Journey by Kayak* (2011 edition).

5 Alastair Dunnett, *Quest by Canoe* (1950), later republished under various titles including *The Canoe Boys* (1995).

6 One excellent collection of oral histories that nonetheless reinforces these categories is Tony Parker, *Lighthouse* (1975).

7 Colin Breen, 'The Maritime Cultural Landscape in Medieval Ireland' in *Gaelic Ireland* (2010).

8 John de Courcy Ireland (ed.), *Ireland and the Sea* (1983).

9 A classic text of Irish ethnology sets up the structures of this society in detail: Robin Fox, *The Tory Islanders: A People of the Celtic Fringe* (1995).

10 David Lloyd, *Irish Times: Temporalities of Modernity* (2008).

11 Pat Conaghan, *The Zulu Fishermen: Forgotten Pioneers of Donegal's First Fishing Industry* (2003), 6.

12 Lillis Ó Laoire, *On a Rock in the Middle of the Ocean* (2005).

13 Ariella Azoulay, 'Potential History: Thinking Through Violence', *Critical Inquiry* (2013).

CONNACHT

1 Sinead Morrissey, 'Achill, 1985' (1996). See also Kathryn Kirkpatrick and Borbála Faragó (eds), *Animals in Irish Literature and Culture* (2015).

2 Robert Harbison, *Eccentric Spaces* (2000).

3 Tim Robinson, *Setting Foot on the Shores of Connemara* (1997).

4 Nessa Cronin, 'Ground Truths: Deep Mapping Communities in the West of Ireland' in S. N. Mayer, T. Lynch, D. Wall and O. A. Weltzein (eds), *Thinking Continental: Writing the Planet One Place at a Time* (1997).

5 Seamas O'Cathain and Patrick O'Hanagan, *The Living Landscape, Kilgalligan, Erris, County Mayo* (1975).

6 Lorna Siggins, *Once Upon a Time in the West: The Corrib Gas Controversy* (2010) is an insightful account of events in Erris; see also *Our Story: The Rossport 5* (2008).

7 Michael Longley, 'The Wren' (2011).

8 A series of excellent appreciations of Robinson's life and work can be found in D. Gladwin and C. Cusick (eds), *Unfolding Irish Landscapes: Tim Robinson, Culture and Landscape* (2016).

9 Pippa Marland, *Ecocriticism and the Island: Readings from the British-Irish Archipelago* (2019).

10 Tim Robinson, interview with David Ward collected in the Folding Landscapes submission to the Ford European Conservation Award, Tim Robinson Archive, NUI Galway.

11 Dorinda Outram, 'The History of Natural History: Grand History or Local Lore?' in J. W. Foster, *Nature in Ireland: A Scientific and Cultural History* (1997).

12 Richard Pine, 'The Cartography of the Soul', *Irish Literary Supplement* (1987).

13 Tim Robinson, *Connemara: the Last Pool of Darkness* (2008).

14 Tim Robinson, *Stones of Aran: Pilgrimage* (1986).

15 Just as for Blasket, there is an extraordinary range of literature from and about Aran; Mairead Conneely, *Between Two Shores: Writing the Aran Islands 1890–1980* (2011) provides a good introduction.

MUNSTER

1 Fiona Green, 'Your Trouble is Their Trouble: Marianne Moore, Maria Edgeworth and Ireland', *Symbiosis* (1997).

2 Nessa Cronin, 'Maude Delap's Domestic Science: Island Spaces and Gendered Fieldwork in Irish Natural History' in N. Allen, N. Groom and J. Smith (eds), *Coastal Works: Cultures of the Atlantic Edge* (2016).

3 Geoffrey Moorhouse, *Sun Dancing: A Medieval Vision* (1998).

BARDSEY TO THE BRISTOL CHANNEL

1 Like most towns on the Atlantic coastlines, hefty tomes of local history detail the lost glories of each of these places, Lewis Lloyd, *Pwllheli: Port and Mart of Llyn* (1991) providing particularly rich coverage of a town transformed by the transition from maritime industry to tourism.

2 Peter Brown, *The Rise of Western Christendom* (1996); Brendan Smith, *Colonisation and Conquest in Medieval Ireland* (1999) and Brendan Smith (ed.), *Britain and Ireland, 900–1300: Insular Responses to European Change* (1999); these kind of phrases echo an earlier tradition in, for instance, Halford Mackinder, *Britain and the British Seas* (1902). See also Sebastian Sibecki, *The Sea and Medieval English Literature* (2008).

3 Edwardes was already famous at this point for his travel writing concerning islands, although he was associated with books about rather warmer climes: *Letters from Crete* (1887) and *Rides and Studies in the Canary Islands* (1888).

4 Griffith's journalistic output tended to be associated with the rural pursuits of the champagne-swilling set to which he most definitely belonged, including titles such as 'Truffle-hunting with pigs and dogs' and 'A Wedding Tour in a Balloon'.

5 R. S. Thomas and Peter Hope Jones, *Between Sea and Sky* (2000).

6 R. S. Thomas, 'The Island' (1972).

7 In September 2018, the Porters left Bardsey for Rhiw (overlooking the island on the mainland); in their eleven years on the island they set high standards of stewardship and hospitality for whoever follows.

8 Key among Evans' Bardsey writings are a prose collection, with photographs by Wolf Marloh, *Bardsey* (2008) and the poetry collection *Island of Dark Horses* (1995).

9 Pippa Marland, 'Island of the Dead: Composting Twenty-Thousand Saints on Bardsey Island', *Green Letters* (2014).

CORNWALL

1 W. S. Graham, *The Nightfisherman: Selected Letters of W. S. Graham* (1999).

2 David Whittaker, *Give Me Your Painting Hand: W. S. Graham and Cornwall* (2015); see also Matthew Francis, *Where the People Are: Language and Community in the Poetry of W. S. Graham* (2004).

3 *Winter Love* might seem to be the odd one out among these poems, rooted firmly in the Mediterranean rather than northern Europe; but its themes

of the poet, as Helen of Troy, as an island in a sea of indifference and solitude, and its representation of Ezra Pound as a journeying Odysseus, all in an aesthetic shaped more by the war years than the decade of publication, means it echoes the atmosphere of the other works mentioned here. Pound's own adaptation of the Anglo-Saxon poem *The Seafarer* (1911) is a much earlier example of similar themes at work.

4 Andrew Lanyon, *Portreath* (1993) and *St Ives* (2003) are rich local studies, among several texts more directly devoted to Peter Lanyon's art.

5 Chris Stephens, *Peter Lanyon: At the Edge of Landscape* (2000).

6 Rosalind Krauss, *The Optical Unconscious* (1993).

7 Peter Lanyon, 1964 lecture to the British Council in Prague.

8 O. G. S. Crawford, 'Lyonesse', *Antiquity* (1927).

THE VIEW FROM THE SEA

1 The particular phrasing here is from a speech given by Mary Robinson, president of Ireland, on Skye in 1997; Robinson was one of the few politicians who pushed the common cause of Irish and Scottish languages. For that reason she is something of a hero among Scottish Gaels.

2 For fuller exposition of these themes, see Keith Robbins, *Nineteenth-Century Britain: Integration and Diversity* (1988) and James Vernon, *Distant Strangers: How Britain Became Modern* (2014).

3 Most debate of Cunliffe's work was among archaeologists or medieval and early-modern historians. Tellingly, the scholars of the modern world who did engage closely with the work were new formations of scholars in Atlantic regions: when the Irish Studies department at the National University of Ireland, Galway, was founded, for instance, it invited Cunliffe to give its inaugural lecture, seeing his work as inspirational for the vision of Ireland-in-the-world they hoped to foster.

4 See, for instance, David Armitage, Alison Bashford and Sujit Sivasundaram, *Oceanic Histories* (Cambridge, 2018).

5 Calls for a new materialism and conceptions of a 'flat social', which doesn't

privilege human agency, run through many current movements in critical theory, from actor network theory to object oriented ontology. In the context of oceans and ecology, the work of Jane Bennett, especially *Vibrant Matter: A Political Ecology of Things* (2009), and Timothy Morton, including *Hyperobjects: Philosophy and Ecology after the End of the World* (2013) are significant contributions. Ecologically, all owe a great deal to Aldo Leopold whose 'land ethic' insisted that, since all plants and animals (including humans) sink or swim together in our fragile ecosystem, all are entitled to citizenship as equal members of the biotic community. The radical work of Donna Haraway and Jason Moore, perhaps best approached through collected editions such as Jason W. Moore (ed.), *Anthropocene or Capitalocene?* (Oakland, CA, 2016) provides further evidence for the importance of these approaches. Historians have made some headway in adapting their ideas, but much less than literary scholars, geographers or social scientists; Timothy Mitchell's 'Can the Mosquito Speak?' remains one of the few convincing historical experiments towards this goal; in relation to this book's themes, the closest thing yet written is perhaps Marcus Rediker, 'History from the Below the Water Line: Sharks and the Atlantic Slave Trade', *Atlantic History* (2008).

6 There have been several calls to end this situation and rethink the skill sets involved in historical practice, such as Mark Levene, 'Climate Blues: Or How Awareness of the Human End might Re-instil Ethical Purpose to the Writing of History', *Environmental Humanities* (2013), and Daniel Smail, *On Deep History and the Brain* (2007).

7 For the beginning of a new, more concerted, attempt to write ocean histories, see the volume mentioned above, David Armitage, Alison Bashford and Sujit Sivasundaram, *Oceanic Histories* (Cambridge, 2018); the work of Sverker Sörlin, on the Arctic ocean, is a particularly insightful contribution combining multiple kinds of history into a genuine study of ocean, not just the maritime movement of people. For the beginnings of a new coastal history movement, see David Worthington (ed.), *The New Coastal History: Cultural and Environment Perspectives from Scotland and Beyond* (2017); Isaac

Land, 'Tidal Waves: The New Coastal History', *Journal of Social History* (2007); and John Gillis, *The Human Shore: Seacoasts in History* (2012). The historical anthropologies of islands that produced classic 1980s texts such as Greg Dening, *Islands and Beaches: Discourse on a Silent Land, Marquesas, 1774–1880* (1980), Marshall Sahlins, *Islands of History* (1985), as well as the early work of Epeli Hau'ofa, have also been revived in work such as Joshua Nash (ed.), *Aquapelago Anthology* (2016), a special edition of the journal *Shima*. When this is combined with intense new interest in Kamau Brathwaite's 'tidalectics' in literary scholarship, leading to publications such as Stefanie Hessler (ed.), *Tidalectics: Imagining an Oceanic Worldview through Art and Science* (2018), there is clearly appetite and potential – even if little of it has yet been achieved – for new coastal conceptions of history.

8 For two calls to arms channelling Carson's ideas see Elyssa Back and J. Baird Callicott, 'The Conceptual Foundations of Rachel Carson's Sea Ethic' in Lisa Sideris and Kathleen Moore (eds), *Rachel Carson: Legacy and Challenge* (2008), and Susan Power Bratton, 'Thinking like a Mackerel: Rachel Carson's *Under the Sea Wind* as a Source for a Trans-Ecotonal Sea Ethic', *Ethics and the Environment* (2004). See also J. Baird Callicott, 'Whaling in Sand County: A Dialectical Hunt for Land-ethical Answers about the Morality of Minke-whale Catching' in Jon Wetlesen (ed.), *Likeverd og Forskjell: En Etisk Intuisjon og dens Grenser* (1997).

9 The ingredients for a more inventive and engaged history, which takes seriously the need to think outside the post-Enlightenment urban ethos of the present, are all in place. Greg Anderson, 'Retrieving the Lost Worlds of the Past: The Case for an Ontological Turn', *American Historical Review* (2015) proposes a new vision for how historians relate to their subjects, which, in combination with the 'potential histories' of Lloyd and Ariella Azoulay, provides a method for giving past outlooks on the world great power in the present. Joan Wallach Scott, 'Storytelling', *History and Theory* (2011); Christine Stansell, 'Dreams', *History Workshop Journal* (2006); Caroline Walker Bynum, 'Wonder', *American Historical Review* (1997); and Robert Rosenstone, 'Space for the Bird to Fly' in S. Morgan, A. Munslow

and K. Jenkins (eds), *Manifestos for History* (2007) provide strong cases for the intellectual necessity of the literary creativity that can give those past perspectives real immediacy. Unfortunately, cross-fertilisation between these cultural histories and environmental history is still stymied by a difference in what kinds of historical explanation each finds compelling. There have been attempts by environmental historians to think through the power of narrative, such as William Cronon, 'A Place for Stories: Nature, History and Narrative', *Journal of American History* (1992), though they still tend to read like caricatures of more cultural approaches. History currently feels like a discipline in which all the ingredients are at hand for transcendence beyond nineteenth-century national, Cartesian and anthropocentric frameworks: the creation from these ingredients of new modes of thinking and writing feels overdue but imminent.

INDEX